Natural History Dioramas – Traditional Exhibits for Current Educational Themes

Annette Scheersoi • Sue Dale Tunnicliffe
Editors

Natural History Dioramas – Traditional Exhibits for Current Educational Themes

Science Educational Aspects

Springer

Editors
Annette Scheersoi
University of Bonn
Bonn, Germany

Sue Dale Tunnicliffe
University College London (UCL)
Institute of Education
London, UK

ISBN 978-3-030-00174-2 ISBN 978-3-030-00175-9 (eBook)
https://doi.org/10.1007/978-3-030-00175-9

Library of Congress Control Number: 2018958716

This Springer imprint is published by the registered company Springer Nature Switzerland AG
The registered company address is: Gewerbestrasse 11, 6330 Cham, Switzerland

Foreword

Our last book, *Natural History Dioramas: History, Construction and Educational Role*, was the first book considering that natural history dioramas are one of the most effective museum exhibit genres for the teaching and learning of many aspects of biology. Dioramas have been, hitherto, a rather neglected area of museum exhibits, but they are experiencing a renaissance, and their educational potential in contributing to peoples' understanding of the natural world is becoming more and more obvious. Dioramas have proved, time and again, that they can reach a wide audience and are a vital tool in increasing the public's access to biology knowledge.

People in the museum and education field approached us about writing a second book. This book is the outcome and focuses on the role of dioramas in addressing challenges of the Anthropocene era.

Science education has been a neglected field in many stages of education with the emphasis being on language and numeracy competency development. However, the world is changing and science learning is becoming recognised as a vital component of a learning journey, especially in this present age of realisation that we humans have contributed to many of the issues facing the planet. This era is being recognised as that of the Anthropocene. Sustainable development is vital to the world, and organisations such as the United Nations have issued goals for this. The understanding of the skills of the STEM subjects (science, technology, engineering and maths) is essential in working towards the goals, together with social competencies such as teamwork, communication and flexibility.

Moreover, the importance of preschool and early years of formal schooling as well as the importance of community involvement is increasingly recognised with a vital role in facilitating learning, not only by formal instruction but by offering informal learning opportunities. In visiting natural history museums with dioramas, all visitors can stand and stare; they can observe the 'big picture' of a whole environment of a moment in time. They may notice the interconnectedness between the earth (geology), climate (meteorology) and organisms (biology) and recognise the different habitats and adaptations of the organisms. Dioramas may also engender an emotional response to our planet and connect the visitors to the issues facing our world at the present time.

This book brings together a collection of voices from different fields that are all associated with natural history dioramas: curators, taxidermists, educators and scientists, all using their expertise to discuss relevant issues from many perspectives and angles.

University of Bonn
Bonn, Germany
Annette Scheersoi

University College London (UCL), Institute of Education
London, UK
Sue Dale Tunnicliffe

Contents

Contributors

Marianne Achiam University of Copenhagen, Copenhagen, Denmark

Charlène Bélanger Université du Québec à Montréal, Québec, Canada

Eric Dorfman Carnegie Museum of Natural History, Pittsburgh, PA, USA

Susanna Edvall Swedish Museum of Natural History, Stockholm, Sweden

Jennifer Gadd Oxford Brookes University, Oxford, UK

Rebecca Gazey Museum of Science and Industry, Manchester, UK

Eirini Gkouskou University of East London, London, UK

Eirik Granqvist Borgå, Finland

Kathrin Grotz Botanic Garden and Botanical Museum, Freie Universität Berlin, Berlin, Germany

Karim Hamza Stockholm University, Stockholm, Sweden

Rainer Hutterer Zoologisches Forschungsmuseum Alexander Koenig, Bonn, Germany

Andrew C. Kitchener Department of Natural Sciences, National Museums Scotland, Edinburgh, UK

Michael May Southern University of Denmark, Kolding, Denmark

Debra McGregor Oxford Brookes University, Oxford, UK

Edward Mifsud University of Malta, Msida, Malta

Alexandra Moormann Museum für Naturkunde, Berlin, Germany

Jesús Piqueras Stockholm University, Stockholm, Sweden

Michael J. Reiss University College London, London, UK

Stephen Rogers Carnegie Museum of Natural History, Pittsburgh, PA, USA

Annette Scheersoi University of Bonn, Bonn, Germany

Rebecca Shreckengast Carnegie Museum of Natural History, Pittsburgh, PA, USA

Till Töpfer Zoologisches Forschungsmuseum Alexander Koenig, Bonn, Germany

Cristina A. Trowbridge American Museum of Natural History, New York, USA

Sue Dale Tunnicliffe University College London (UCL), Institute of Education, London, UK

Chapter 1
Introduction: Natural History Dioramas and Science Educational Aspects

Annette Scheersoi and Sue Dale Tunnicliffe

Museums are now regarded as almost the last public space left available to all. They should view themselves as communal meeting places, in which people can even discuss controversial issues (Sharp 2016). Museums can act as facilitators of communication and collaboration between scientists and the general public on issues of the twenty-first century including natural and social sciences (Garthe 2018).

Natural history dioramas, with their taxidermied animals and representations of (authentic) habitats, can help us to visualise the consequences of human activity by addressing environmental issues (Wonders 2016). They provide opportunities for, and often spark debates, to talk about relevant controversial topics such as climate change, global warming, loss of habitat, industrial pollution and the dominance of one particular species, *Homo sapiens*, over many others, in some cases to the point of extinction.

Originally, natural history dioramas were a nineteenth century development, but they have evolved into institutions for both researching biodiversity and more recently focusing on the changes during the Anthropocene era (Crutzen 2002) – and how such information can be effectively accessed and understood by visitors.

A natural history museum is an essential component within the educational or learning system in a country. Hence, its contents, which are thus subsets of the mission of the museum, are important individual entities within an ecosystem of science learning. Individuals develop their science understanding both in and out of school and such acquisition is life-long. Therefore, there are many constituents in the overall learning community that contribute to such learning opportunities.

A. Scheersoi (✉)
University of Bonn, Bonn, Germany
e-mail: a.scheersoi@uni-bonn.de

S. D. Tunnicliffe
University College London (UCL), Institute of Education, London, UK
e-mail: s.tunnicliffe@ucl.ac.uk

A. Scheersoi, S. D. Tunnicliffe (eds.), *Natural History Dioramas – Traditional Exhibits for Current Educational Themes*,
https://doi.org/10.1007/978-3-030-00175-9_1

Traditional sources of learning are only one source, particularly in science where, as adolescence begins, young learners often loose interest in formal science in schools. Research emerging in the last quarter of the twentieth century, which build on earlier visitor studies, focused on Out-of-school learning. More and more research considers the contributions of places, such as botanic gardens, zoos, farms, (science) museums and other Out-of-the-classroom learning sites, for science education (e.g. Braund and Reiss 2004).

Tal (2016) points out that community based 'non-organisations' activities and other inputs in communities – from community gardens and nature centres to digital media sources of information – are increasingly playing an important role in the learning journey of an individual and that traditional divisions between 'formal' establishments such as schools and 'informal' institutions are not precise. Tal (2016) furthermore raises the issue of pedagogies, in terms of form of interactions between museum educators and the visiting learners, where many museums educators use traditional knowledge transmission models and assume the role of the expert disseminating information to the beginner. The transmission mode of learning requires the receiver to consciously participate in the transmission. Indeed, Vygotsky (1986) suggested that 'any function in a child's cultural development appears twice, or on two planes, first it appears on the social plane and then on the psychological' (p. 16). Following this line of thinking, educators could recognise the importance of social interaction, talk or dialogic dialogue (Alexander 2008), particularly in working with school groups visiting museums in the widest sense, as well as in schools. Museums increasingly recognise that socio-cultural exchanges are both between the visitors with all their pre-visit understanding through which they interpret the exhibits, between generations, social classes, peers of the groups and other visitors, as well as museum personnel.

Museums of all genres have, as one of their missions, to foster understanding of the subject focus of the museum, be it arts or science. However, Anderson (2016) identifies barriers and inhibitors, which exist and render this missions' intent more difficult. He argues that first of all, in the museum itself, educators, scientists, in-house consultants as well as exhibit designers, might hold differing expectations of outcomes. Anderson further asserts that there are mismatches in museum pedagogy, which limit the effectiveness of science museums as cultural resources for learning. These barriers counteract the museums' main goals. He suggests that museums need to recognise the distinct cultures to which museums' visitors belong, in order to understand the way in which they interpret the exhibits. Thus, museums need to identify their audience and tailor their pedagogies to allow for such variation. Museum visitors are recognised as being a widely heterogeneous group, ranging from families with young children, adult groups, leisure visitors to those of formal and adult education groups, particularly in the case of science museums in the broadest sense. One-size initiatives do not 'fit all'. Hence, a progressive understanding of the changing profile of visitors and their interpretation is an essential part for museums in order to develop an accurate view of their audiences.

It has been shown that dioramas provide constructive learning opportunities for a wide range of visitors: If a diorama provides a variety of anchor points it enables visitors to relate their previous experiences and knowledge to the scenes or artefacts presented. It often results in visitors' feelings of enjoyment, involvement, and stimulation, which are the most typical emotional aspects of an interest-based activity (Scheersoi 2015). Tunnicliffe and Scheersoi (2010) maintain that the focus of intervention initiatives should be on accurate minds-on observations, rather than pure physical hands-on manipulation of objects, inviting the observer to ask questions. Facilitators, employed by the museum, or members of a visiting group can encourage minds-on focus at dioramas and, through using appropriate 'talk', assist others in developing understanding.

Dioramas are an established form of exhibit in museums, bearing both cultural and scientific significance, particularly in natural history museums. They are windows into a natural or a human constructed world, depicting the past or the present and sometimes even pointing to the future. Hence, they are fascinating for visitors and many educators consider dioramas to be essential learning tools. They are 'minds-on' exhibits as opposed to 'hands-on' in which the physical interaction frequently becomes the exhibit. Reiss and Tunnicliffe (2011) made the case for biological dioramas but also point out that there is scant literature about such, although it is increasing, e.g. Tunnicliffe and Scheersoi (2015). Another recent diorama book edited by Gall and Trischler (2016) is mainly devoted to other types of dioramas, e.g. modeled scenes in science museums, but also includes some chapters dealing with natural history dioramas.

Dioramas can be accurate life size representation of actual scenes, as for example in the Carl Akeley Hall of Mammals in the American Museum of Natural History in New York, where photographs of a scene portrayed are shown alongside the constructed reality diorama (Quinn 2006). They may be considered as constructs of reality, likened to a photograph, capturing a moment in time (Howie 2015). However, dioramas can also be designed to portray concepts. They might reflect socio-cultural attitudes or the ethos of different countries and can show us how perspectives have changed over time, like the dioramas in Salzburg (Haus der Natur), Austria.

Traditional dioramas are life size but there are small dioramas rather like the size of glass aquaria one knows from homes and laboratories: Examples are presented in the Booth Museum of Natural History in Brighton, which together tell a particular story of the variety of species in Britain. Another example is the Botanical Museum in Berlin where small dioramas represent different vegetation types around the world. Such small-scale conceptual dioramas, with miniatures of the artefacts shown, are referred to as 'Little Landscapes' (Insley 2007).

Originally, life-sized habitat dioramas were designed to provide an authentic replication of the fauna, flora and topography of 'exotic' regions, which only a few people were able to visit. They were the media of their time, informing the populace at large of biodiversity, which was different to that of their native land. Although some collectors were trophy hunters, some possessed altruistic motives and, financing the expeditions by private means or commissions from collectors, collected, preserved and displayed specimens for local people in their neighbourhood to view. Such a person was Major Percy Powell-Cotton. He collected animals in India and Africa, and built a gallery at his Kent home, Quex Park, to display them (Joynes

2016). In an era without television and film, taxidermic animals and their environments as portrayed in natural history dioramas, were a major source of amazement to local people (Joynes 2016). In present times, there are many forms of media depicting such scenes and with the advent of reasonably priced mass air travel people are now able to visit such locations for themselves. Another aspect was the emergence of interactive science centres in the second half of the twentieth century; they became an important place for acquiring science concepts, often through experiential hands-on opportunities. As a consequence, many museums with natural history dioramas regarded them as old fashioned and dismantled them. They now had to add another component to their exhibitory. Natural history museums introduced push button and telephone access exhibits along with computer displays and live shows. So, in the second half of the twentieth century, natural history museums changed from focusing on well-presented biological scenarios, which had succeeded the myriad rows of preserved specimens in cabinets for scientific perusal, to hosting the ‘block buster’. Often these travelled around and an admission charge was made (Radar and Cain 2014), traversing oceans to different continents as do ‘block buster’ art and other cultural exhibitions.

But the twenty-first century is seeing a reappraisal of the importance of natural history dioramas and their contribution to both cognitive learning opportunities and the affective domain; being aware of the tremendous social opportunities museum visits grant to people of all backgrounds. Natural history dioramas are again increasingly recognized as a valued genre of exhibit and as evidence of changes in the environment. They become historical documents, and dioramas which escaped demolition, such as the zoogeographical dioramas of the Hessische Landesmuseum (Darmstadt, Germany) that were originally set up at the beginning of the twentieth century, are being carefully renovated (Munsch et al. 2015). Moreover, new dioramas, such as those telling the story of the history of the fauna and flora of Scotland from the last Ice Age onwards, have been installed (Museum of Scotland, Edinburgh) (Andrew 2018).

Nevertheless, inexperienced museum staff is one of the biggest issues in creating and preserving dioramas for future generations, and can lead to the dioramas’ destruction. Practitioner knowledge and academic theory have to be both joined in contemporary habitat diorama artistry and exhibition planning. The major challenge is to bring together curators, exhibitions designers, taxidermists and model makers, educators and scientists, all using their individual skills and expertise and working hand in hand. In this book, we bring together a collection of voices from such experts in different fields that are all associated with natural history dioramas to discuss relevant issues from many perspectives and angles.

Two books are simultaneously published. While the book at hand is particularly focused with science educational aspects, our second book focuses on socio-cultural issues and the potential of using dioramas to engage various audiences with – and in – contemporary debates and big issues, which society and the natural environment are facing.

In this book the first five chapters consider the role of museum personnel, such as curators, taxidermists, and scientists, in designing and realising particular natural history dioramas.

Rogers, Shreckengast and *Dorfman* from the USA (Carnegie Museum of Natural History, Pittsburgh) summarise the early history of habitat dioramas in the US, highlighting Charles Willson Peale's first displays in the late 1780s and trends in the US to 1900. Taxidermy and habitat publications, exhibitions, and important individuals are described. Additionally, the authors explore the current conditions of habitat diorama artistry and exhibition planning in a selection of today's US natural history museums.

Kitchener from Scotland (National Museum of Scotland, Edinburgh) describes the background and development of a new diorama in his museum as part of the *Beginnings* gallery, which presents changes in the geology and wildlife of Scotland over that last three billion years. The aim of the diorama was to tell a story about changing habitats and their associated fauna and flora since the end of the last Ice Age almost 12,000 years ago. The importance of the diorama in interpreting contemporary climate change and ecology is also discussed. Finally, the chapter presents a preliminary evaluation, which gives insights into how visitors interact with and respond to the diorama.

Granqvist, a taxidermist from Finland, contributes a personal account of how natural history dioramas were designed and constructed in Helsinki in the third quarter of the twentieth century in Europe. It reviews some reasons for using dioramas in a museum and describes how to build and maintain a diorama. The chapter points out that these dioramas are replications of actual habitats with their fauna and flora and valuable records of changes in the last century.

Hutterer and *Töpfer* from Germany (Museum Alexander Koenig, Bonn) show how birds played a crucial role in the development of museum displays and their most sophisticated examples, the habitat dioramas. Breeding colonies of marine birds were among their first subjects and served as role models for similar presentations in museums around the world. They review important examples of this type of diorama, document the history and analyse the design of a still existing bird colony diorama from the early 1930s in the Museum Koenig in Bonn from an educational point of view. Based on this example they present different ways of extracting interesting stories and entertaining information for different audiences. They show that the seabird diorama provides multifaceted potential for different biological and historical narratives.

Grotz from the Botanical Museum in Berlin (Germany) writes about their 16 dioramas, unique in their exclusive focus on plants, their diversity in specific habitats around the world, as well as their downscaled size. These botanical dioramas from the 1950s and 1960s both challenge and enrich the museum's efforts to communicate vegetational changes and threats. The author both focuses on the genesis and scientific background of these elaborate objects, their potential and limitations as contemporary agents to transfer science knowledge.

The second section of this book focuses on different theoretical aspects of learning with natural history dioramas.

Moormann from Germany (Museum für Naturkunde, Berlin) and *Bélanger* from Canada (Université du Quebec, Montréal) discuss the educational role of dioramas as models. Since museum dioramas can be understood as simplified external representations of scientific models, they infer that dioramas have the potential to promote model-based learning. By engaging with dioramas, the visitors' mental models become more complex and gradually align with the scientific models from which they were derived.

May and *Achiam* from Denmark (Southern University of Denmark and University of Copenhagen) provide a theoretical foundation, based on Gestalt psychology and cognitive linguistics, for analysing the museum visitors' construction of meaning when viewing natural history dioramas. This process of meaning construction is presupposed in any discussion about informal or formal learning from dioramas. Empirical studies of visitor behaviour and verbal reports can indicate how the visual opportunities and constraints of the diorama might have influenced the experience of visitors. However, a meaningful experience is actually the end product of a mental construction of meaning that starts with Gestalt perception and ends in high-level constructions such as narratives and discourse perspectives.

Mifsud from Malta (University of Malta) points out that museum artefacts and settings are subject to viewer interpretation. He explores the elements involved in interpreting natural history dioramas and proposes a novel interpretation model for museum settings based on Engestrom's Activity System. In this context, a natural history diorama has the potential to be a unique model in biological learning.

The third section consists of four chapters, which discuss science learning activities involving dioramas.

McGregor and *Gadd* from the UK (Oxford Brookes University) explore how „Do-it-yourself" dioramas, constructed in shoe boxes, can contribute to teaching and learning. The evidence from constructing and reviewing home-made exhibits indicates that they can offer unique opportunities to better understand concepts related to natural history.

Trowbridge from the American Museum of Natural History in New York (USA) shows that natural history dioramas are generative visual stimuli, which afford various entry points for people to engage in science. Observations from work with new science teachers who graduated from the Museum's Masters of Arts in Teaching program and participated in the Museum's New Teacher Induction activities, suggest that the use of modified Visual Thinking Strategies (VTS) with natural history dioramas have a positive impact on teacher learning and reflection.

Tunnicliffe from University College London Institute of Education, *Gazey* from the Powell Cotton Museum (Kent) and *Gkouskou* from University of East London, all three from the UK, focus on aspects of physical science in action shown in the animals featured in dioramas. Basic ideas such as shape, size, colours and patterns, as well as basic numeracy for earliest learners were explored. Older primary children can explore basic forces, balance and centre of mass, as well as sound and light. Effective learning strategies involve knowing the foundation knowledge that the children possess.

Piqueras, *Hamza* (both from Stockholm University) and *Edvall* from the Swedish Museum of Natural History in Stockholm present a detailed analysis of students' moment-by-moment learning during a science teaching activity involving dioramas in a natural history museum. The content and the directions learning took in response to what students noticed in the diorama revealed that certain details and elements of the diorama helped significantly in their process of meaning-making. Interestingly, one of the most discussed features of the diorama in the students' reflections was not related to the main pedagogical intention of the diorama.

This book is concluded with a chapter from *Reiss* (University College London Institute of Education, UK) reflecting that natural history dioramas have long been identified as potentially valuable sites of learning. He examines what the purposes of science education might be and then argues that the other chapters in this book show that natural history dioramas do indeed have considerable potential to advance the learning of science. Dioramas are engaging for many visitors, whatever their age and prior knowledge and experience, and enable them to construct narratives about what is happening in the diorama.

The second book (Natural History Dioramas – Traditional Exhibits for Current Educational Themes, *Socio-cultural Aspects*) includes chapters from a wide range of educational practitioners. They discuss the role of dioramas not only in contributing to scientific understanding and cultural awareness, but also in reconnecting wide audiences to the natural world and thereby contributing to the well-being of societies.

References

Alexander, R. (2008). *Towards dialogic teaching: Rethinking classroom talk* (4th ed.). Dialogos: York.

Anderson, D. (2016). *Innovations in science education. Research in practice.* Key note talk given at EASE 2016 Tokyo, Strengthening International Collaboration. Tokyo, August 2016.

Andrew, C. K. (2018). Chapter 3: Developing a modern diorama: Changing habitats in Scotland since the end of the Last Ice Age. In: A. Scheersoi, & S. D. Tunnicliffe (Eds.), *Natural history dioramas – Traditional exhibits for current educational themes* (Science Educational Aspects). Cham: Springer.

Braund, M., & Reiss, M. J. (2004). *Learning science outside the classroom.* London: Routledge.

Crutzen, P. J. (2002). The geology of mankind. *Nature, 415*, 23.

Gall, A., & Trischler, H. (Eds.). (2016). *Szenerien und Illusion: Geschichte, Varianten und Potenziale von Museumsdioramen.* Göttingen: Wallstein.

Garthe, C. J. (2018). The natural futures museum: Interactivity and participation as key instruments for engaging audiences. In E. Dorfman (Ed.), *The future of natural history museums, ICOM advances in museum research series* (pp. 140–154). London/New York: Routledge.

Howie, G. (2015). Dioramas as constructs of reality: Art, photography, and the discursive Space. In S. D. Tunnicliffe & A. Scheersoi (Eds.), *Natural history dioramas, construction and education anal role* (pp. 39–65). Dordrecht: Springer.

Insley, J. (2007). Setting the scene. *Museums Journal, 107*(2), 33–35.

Joynes, A. (2016). *Tracking the Major. Sketches from the Powell-Cotton Museum.* Canterbury: Mickle Print Ltd.

Munsch, M., Schmiese, H., Angelov, A., Riedel, G., & Köhler, J. (2015). Conservative restoration and reconstruction of historical natural history dioramas. In S. D. Tunnicliffe & A. Scheersoi (Eds.), *Natural history dioramas. History, construction and educational role* (pp. 115–130). Dordrecht: Springer.

Quinn, S. C. (2006). *Windows in nature: The great habitat dioramas of the American Museum of Natural History*. New York: Abrams.

Radar, K. A., & Cain, V. E. M. (2014). *Life on display. Revolutionizing U.S. Museums of Science and Natural History in the twentieth century*. Chicago: University of Chicago Press.

Reiss, M., & Tunnicliffe, S. D. (2011). Dioramas as depictions of Reality and opportunities for learning biology. *Curator, 54*, 447–459.

Scheersoi, A. (2015). Catching the visitor's interest. In S. D. Tunnicliffe & A. Scheersoi (Eds.), *Natural history dioramas. History, construction and educational role* (pp. 145–160). Dordrecht: Springer.

Sharp, R. (2016). Action points. How museums are addressing social and political isssues. *Museums Journal, 11*(2016), 29–33.

Tal, T. (2016). *Global perspectives on trends in informal science education research*. Symposium talk given at NARST 2016 conference, 15th April, Baltimore, USA.

Tunnicliffe, S. D., & Scheersoi, A. (2010). Dusty relics or essential tools for communicating biology? In A. Filippoupoliti (Ed.), *Science exhibitions: Communication and evaluation* (pp. 186–216). Edinburgh: MuseumsEtc.

Tunnicliffe, S. D., & Scheersoi, A. (2015). *Natural history dioramas. History, construction and educational role*. Dordrecht: Springer.

Vygotsky, L. (1986). *Thought and language*. Cambridge: Harvard University Press.

Wonders, K. (2016). The habitat diorama phenomenon. In A. Gall & H. Trischler (Eds.), *Szenerien und Illusion. Geschichte, Varianten und Potenzial von Museumsdioramen* (pp. 286–318). Göttingen: Wallstein.

Annette Scheersoi is a Professor in Biology Education at the University of Bonn (D). Her doctorate was on museum's exhibition concepts and the use of different media for communicating biological knowledge. Since 2005 she is a biology teacher-trainer and researcher, first at Frankfurt University, then at the University of Cologne and now at Bonn's university. Her research on biology learning focuses on interest development in out-of-school learning environments. Annette also develops learning materials for schools as well as for museums, zoos and botanic gardens.

Sue Dale Tunnicliffe is a Reader in Science Education at UCL Institute of Education, University College London. She holds a bachelors degree from Westfield College, University of London in Zoology and trained as a secondary biology teacher at the university's Institute of Education. After teaching in grammar schools for a few years she had children and, wanting to find out more about how young children learnt science, particularly biology, she entered primary school teaching and is now interested in pre school learning, as well as animals as exhibits. She set up and ran a new Primary Science and Design Technology advisory team for London Borough of Richmond and then became Head of Education at the Zoological Society of London. She has worked at the BBC and in a cultural museum. Her doctorate was from King's College, London 'Talking about animals: conversations of children in zoos, a museum and a farm". She has published widely.

Part I
Dioramas as Witnesses of the Past and Evidence of Change

Chapter 2
Origins and Contemporary Status of Habitat Dioramas in the United States

Stephen Rogers, Rebecca Shreckengast, and Eric Dorfman

2.1 Introduction

Whether they are in or out of fashion, habitat dioramas have been a mainstay of natural history exhibition in United States museums since the early 1900s. Before there were habitat dioramas, there was taxidermy. An American taxidermist and museum innovator, Charles Willson Peale, first showcased taxidermy specimens within a recreated, natural environment 225 years ago. Since his innovation the primary function of taxidermy and habitat dioramas has been disputed. Are they for decoration and entertainment, scientific study, or primarily for education—perhaps all three. Since Peale's first efforts, habitat dioramas have evolved; terms like "groupings," "foreground," and "background painting" emerged and influenced composition and style. Although most taxidermy techniques came from European teachings and publications, the innovation and commercial publishing success of American taxidermists during a peak museum construction period created the great legacy of habitat dioramas in United States natural history museums.

Taxidermy is the art of preparing vertebrate specimens for long-term preservation. Taxidermists skin the animal and remove parts that could break down inside the skin; they then replace those parts with materials that will not decompose. Their aim is to recreate lifelike or animated facsimiles of what was once alive. In its early history in Europe the function of taxidermy was primarily decorative, but as science advanced into the mainstream, by the 1750s preserved specimens became necessary to support rigorous study.

S. Rogers (✉) · R. Shreckengast · E. Dorfman
Carnegie Museum of Natural History, Pittsburgh, PA, USA
e-mail: RogersS@CarnegieMNH.Org

A. Scheersoi, S. D. Tunnicliffe (eds.), *Natural History Dioramas – Traditional Exhibits for Current Educational Themes*,
https://doi.org/10.1007/978-3-030-00175-9_2

The art of taxidermy is typically learned in one of two ways: by working directly with a person who has mastered the art form or by reading descriptive literature and then practicing until those skills are acquired. Within the United States, the development of taxidermy, and then ultimately how the specimens were displayed, was a mixture of both methods. The following chapter will explore the major milestones that led to the advancement of taxidermy and the early development of the habitat diorama within the United States.

2.2 Peale's Museum

Taxidermy came of age in the United States in the mid-1780s, partly due to information gained from European publications and knowledge exchanged through travel. At that time, Benjamin Franklin returned to Philadelphia from France and again assumed the helm of the American Philosophical Society. He embraced the idea of a museum proposed by fellow Philadelphian, Charles Willson Peale, a skilled artist who wished to build a repository of natural curiosities. Franklin had seen many museums in Europe and realized that in order to support the Society's mantra "Promote Useful Knowledge," Peale's repository would need to display natural history specimens. In 1784, Franklin lent Peale publications on preservation by Louis Marie Jean Daubenton of France, and John Coakley Lettsom from England (Sellers 1980). Daubenton's text, as were many of the time, primarily concerned itself with the preservation of specimens for shipping back to France (Farber 1977). Lettsom's publication included very little actual information on taxidermy.

Peale used his ingenuity to fill in the knowledge gaps from these texts, which were lacking in many details, and his inventiveness to develop preservation skills. He started creating natural history specimens for his residence, which housed his first attempt at a museum. Peale also accepted donations of preserved specimens and began exchanging specimens with fellow enthusiasts. Previously, Peale had built, adjacent to his residence, a room with a skylight and, by late 1786, he designed an entirely novel technique of displaying his preserved specimens. Rembrandt Peale, one of Peale's sons (Sellers 1980), recorded a description of the exhibit area by a visitor to the museum from Massachusetts written sometime between 1786 and 1790:

> At the opposite end, under a small gallery, his natural curiosities were arranged in a most romantic and amusing manner. There was a mound of earth, considerably raised and covered with green turf, from which a number of trees ascended and branched out in different directions. On the declivity of this mound was a small thicket, and just below it an artificial pond; on the other side a number of large and small rocks of different kinds, collected from different parts of the world and represented in the rude state in which they were generally found. At the foot of the mound were holes dug and earth thrown up, to show the different kinds of clay, ochre, coal, marl, etc. which he had collected from different parts; also, various ores and minerals. Around the pond was a beach, on which was exhibited an assortment of shells of different kinds, turtles, frogs, toads, lizards, water snakes, etc. In the pond was a collection of fish with their skins stuffed, water fowls, such as the different species of geese, ducks, cranes, herons, etc.; all having appearance of life, for their skins were

admirably preserved. On the mound were those birds which commonly walk on the ground, as the partridge, quail, heath-hen, etc.; also different kinds of wild animals,—bear, deer, leopard, tiger, wild-cat, fox, raccoon, rabbit, squirrel, etc. In the thicket and among the rocks, land-snakes, rattle snakes of an enormous size, black, glass, striped, and a number of other snakes. The boughs of the trees were loaded with birds, some of almost every species in America, and many exotics. In short, it is not in my power to give a particular account of the numerous species of fossils and animals, but only their general arrangement. What heightened the view of this singular collection was that they were all real, either their substance or their skins finely preserved…(pp 27–28).

This newly developed style of exhibiting large groups of taxidermy specimens in recreated, natural habitats lasted only a few years at Peale's museum. It is likely he abandoned the display style because the uncased specimens were subject to damage by insects, patrons, and ultraviolet rays from the skylight. As his collection expanded, Peale outgrew the space and moved to a gallery in the American Philosophical Society. Later, when he moved his museum to Independence Hall in 1802, he built floor to ceiling, stacked, side-by-side cases to display his mounted specimens. European display fashions were similar, as seen in the paper by Wonders (1993a), but Peale innovated the design by adding preserved vegetation and painted backgrounds to these cases "the insides of which are painted appropriate scenery, mountains, plains or waters, the birds being placed on branches or artificial rocks" (Sellers 1980). Often the background paintings were of specific locations, sometimes replicating the exact spot where the specimen was collected from. Peale's cases were, by all accounts, the first habitat dioramas in the United States as well as the entire world.

2.3 Publications on Taxidermy in the United States

In 1800 there were very few practitioners of taxidermy within the United States. Enthusiasts of this art visited with Peale or one of his sons, who were also taxidermists, in order to learn proper preparation. Still, scant information was available in English regarding preparation. In 1803, the word *taxidermie* was coined by Louis Dufresne, who worked at the Muséum national d'histoire naturelle in Paris. Dufresne contributed pages 507–565 in the *Dictionnaire d'Histoire Naturelle,* which was by far the most detailed information on taxidermy written to that time. A second expanded edition of the *Dictionnaire* appeared in 1819, and then Dufresne published a stand-alone book on taxidermy (Dufresne 1820) that was translated into English by Sarah Bowdich (Anon. 1820; Beaver 1999) and published in multiple editions until 1849. In this text, no reference was made to "dioramas" or "groupings" with only a small passage describing placing specimens in a box: "the interior of this is furnished with white paper, well pasted; in front is a groove, to receive a glass, fastened by four points and closed with putty."

The first comprehensive United States publication detailing taxidermy was by Samuel Kettle (1831). Kettle's book was mostly a translation of a text by Boitard (1825) who had freely borrowed information from the Dufresne work (1820). Kettle rearranged text and added some reference to taxidermy and taxidermists in the United States. The section on foreground work was embellished, perhaps because it was deemed important. Kettle's passage covers groupings and foreground:

> Of Groupings
>
> By this term, we understand the arrangement of two or more subjects together, so as to represent an action; such as—a falcon grasping a dove within his talons,—a partridge covering her brood with her wings, in defending them from the attacks of a weasel or hawk,—a pair of doves perching and billing upon a rose-bush,—mockingbird-bird essaying a vain defence against the fangs of a snake, who is thrusting his head into her nest &c. Compositions of this sort form a striking and interesting kind of picture, when they are arranged with taste and skill. To this end, the operator should be able to bestow upon each individual, the attitude and expression fitted to denote the particular emotion which he is imagined to feel,—as rage, grief, ferocity, love, &c. ... (pp. 145) [The author then describes extracts from Boitard's work on attitude and descriptions that birds make during various "passions" of their life].
>
> The branches, upon which birds are generally placed, are gathered upon the skirts of woody spots; they are commonly the limbs of plum-trees, which have been stunted by the bite of cattle, and become covered with white and yellow lichens; the branch, fixed in a cabinet, is adorned with artificial flowers and leaves fastened on with wire. To imitate the various sorts of masses, lichens and short grasses, it is customary to use the fine shavings of horn made by turners; these are colored according to fancy, and sifted over the branch—which has previously been coated with glue or paste. An imitation of rock is effected by brown pasteboard, wetted in this paste, molded to the proper shape and covered with fine sand. Earth is imitated with sand, gravel, coffee ground, &c.
>
> Beyond these general ideas, the operator must be left to the guidance of his own taste and ingenuity; groups of the above description possess value both for use and ornament; when arranged with skill and effect, they constitute alike objects of scientific study and elegant taste (pp 148).

Two years later, a book written by Captain Thomas Brown in Edinburgh (Brown 1833) became the most widespread treatise on taxidermy in the nineteenth century: *The Taxidermist's Manual.* First published in Glasgow, this publication was again based primarily on work by Dufresne but also injected passages from Charles Waterton and other practitioners. This popular text went into at least 28 exact reprints in England into the 1880s and was reprinted as an inexpensive paperback in the United States. No description of foreground or background painting was included in the entire text.

About 1870, an America Revised Edition from the Twentieth English edition of Brown's work was modified and published with officers of the Smithsonian Institution (Anon. 1870). This work had extensive details about the creation of cases and habitats for mounted taxidermy. After suggesting bell jars as a possible way to preserve bird specimens (because the price was "now so reasonable"), the text goes into a description of cases:

> Wooden cases should be made as slight (in thickness) as is consistent with firmness; well seasoned white pine is best; and the case should be formed to back, top and bottom, open at the front and sides... having the case prepared, it should be papered with strong manila paper on the top and back within, and when paste is dry, washed over carefully with size and whitening, tinted with a little stone-blue; some add touches of white subsequently to represent clouds,... some paste landscape on the back. The bird to be placed in this case is

either perching, standing, or flying... fastened to the bottom of the case, either by screwing from below, from above, or gluing... The bird being fixed, the next thing is the decorating or "weeding", as it is technically called... tufts of grass... the whole bottom should be carefully glued over with thin glue... fine-sifted sand or gravel should be sifted over it.... Moss [and]... plants with water-color with gum will be used with advantage and looks less artificial than oil paint... ferns look very pretty... if [one wishes to have] a lump of earth, or an apparent bank, a piece of brown paper bent to the requisite shape, and glued over and covered with sifted sand or gravel has a very good effect. Regard should also be made to the season at which the bird is usually seen. For instance, summer birds are of course surrounded by green and living objects, but autumn or winter visitants by decaying or dead herbage. It has often been made an experiment to represent snow, but it is difficult to obtain anything white enough, and at the same time of a crystalline character, which, of course, it should be. Potato farina, nicely dried, mixed with Epsom salts pounded very fine, does not make a bad substitute... a very white mineral powder mingled with pounded glass is perhaps best... to represent water a small piece of looking-glass surrounded by moss, &c. answers very well (pp 46–48).

The Smithsonian text describes a three-sided case that varied from the European model of a case open only on the front. If clouds and sky, as well as a painted landscape were completed, and the foreground built, this is exactly the description of a classic habitat diorama. An example can be seen in Fig. 2.1.

Fig. 2.1 Here we show a photograph of a small grouping of birds at Carnegie Museum of Natural History, open on three sides, modeled after this described approach with lichens and crushed glass and a mirror "looking-glass" placed to represent water. It is overstuffed with a wide mix of specimens from many continents, but conforms to the design described in the American Revised Edition of Brown (Anon. 1870)

The rationale for building the habitat group as described could have come from recollections of the Peale museum. However it came about, from that time forward publications in the United States often describe extensive enhancement to the plain taxidermy specimen which, up to that time, was commonly placed upon a T-perch or on a piece of finished wood. Museum display, which had been for scientific study and classification purposes, was evolving into a new presentation style that was both educational and dramatic. These displays told stories about the subjects in their environment.

2.4 Analysis of Trends in Publication

One means of measuring the growing importance of taxidermy and the importance of habitat materials and accessories is by quantifying United States publications from 1800 to 1900 that give directions for performing taxidermy or discuss progress in taxidermy. For inclusion in the analysis for this chapter, the publication must have had at least ten pages of text and be published by an American, even if it is primarily written as or translated from an earlier, European work. Most taxidermy publications throughout the world make liberal use of earlier works and usually do not credit the original that forms much of the information. Many smaller publications on taxidermy exist, but these are usually brief and inadequately describe display methods, so were not tabulated. A large number of publications describe scientific preparation or field methods without any listing of taxidermy; these texts have not been included. A goodly number of the latter publications can be seen in Rogers et al. (1989).

Each publication was evaluated for its inclusion of habitat or foreground work included in the text or shown graphically. Examining these publications by decade (Fig. 2.2) shows a trend in both the number of publications and those that consider

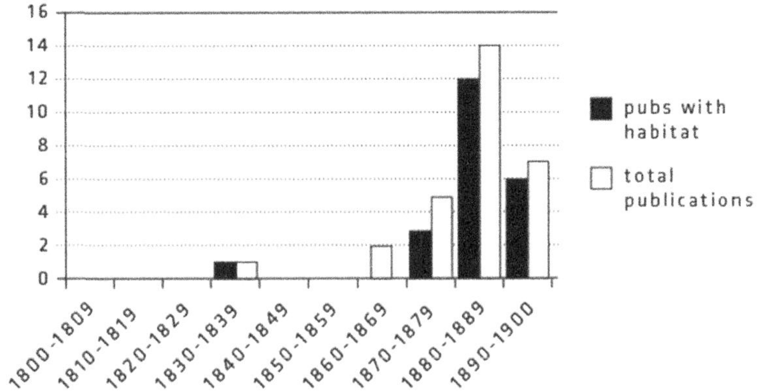

Fig. 2.2 Number of United States taxidermy publications by decade and portion including details on habitat construction

real or artificial branches or stumps, or artificial vegetation and groundwork vs. a T-perch or finished piece of wood as a base (Citations used to construct the graph: Allen 1885; Anon 1870, 1887, 1890; Batty 1880; Coues 1874; Davidson 1885; Davie 1882, 1894, Harding 1875; Holder 1861; Hornaday 1891; Kettell 1831; Kingley 1882; Manton 1882; Maynard 1873, 1883; McConnaughay 1898; North 1882; Reed 1990; Rowley 1898; Shufeldt 1894; Society of American Taxidermists 1881, 1882, 1884; Sylvester 1865; Walker 1870; Ward 1881–1886; Warren 1896; Webster 1885–1886).

2.5 Early Habitat Groups in the United States, 1850–1876

In England, an era of widespread interest in taxidermy commenced with the "Great Exhibition" in 1851 where a number of European taxidermists exhibited preserved specimens and habitat groups in Hyde Park, London. The United States answered with a similar "World's Fair" exhibition that opened in 1853. One million, one-hundred thousand people saw the New York World's Fair exhibition before it concluded in November 1854. Were it not for James A. Hurst, taxidermy might have gone unnoticed at the event. Hurst was the only taxidermist listed in the first catalogue:" Preserved game birds of various countries; preserved Albino and mottled deer; preserved panthers, male and female—James A. Hurst, taxidermist. Old State Hall, Albany New York." In a revised edition, three other taxidermists had joined the exhibition but Hurst created the largest and most impressive display of its kind with additional "preserved owls; ourang-outang [sic]; green monkey; fish and other animals" in his exhibit space (Anon. 1853).

James A. Hurst has been entirely overlooked in major publications detailing the early history of dioramas and habitat groups in the United States. (Shufeldt 1894; Lucas 1914a, b; Hornaday 1922, 1925; Rowley 1927a, b; Webster 1945; Wonders 1993a, b). Robert Hatt (1976) details most of what we know about Hurst drawn from newspaper clippings and the many stereoviews Hurst produced and registered as sets in the 1870s.

Hurst was born in England in 1810 but immigrated early to Canada and then moved to Utica, New York, in 1847 where he established a taxidermy shop. In 1849 he was persuaded to move to Albany, New York, for a part-time position with the Cabinet of Natural History; there, Hurst worked until his death in 1882. During his 33 years in Albany, Hurst operated a private taxidermy business mounting specimens for customers or selling specimens he had acquired. Much of his side work was devoted to creating his own extensive collection of birds and beasts of North America. He used these collections to produce stereoscopic views with dual purpose, "For Schools and Parlor Entertainments" and to supplement his income (Figs. 2.3 and 2.4). Before fast shutter speeds and television made images of wildlife readily available to a large, general audience, Hurst's stereoviews of taxidermy carefully placed in the context of their environment offered the public an innovative way to view a representation of the natural world from the comfort of home.

Fig. 2.3 *The American Wolf* with accessories and painted background; photographed prior to the establishment of Hurst's Free Museum; *Hurst's Stereoscopic Studies in Natural History, For Schools and Entertainment No. 3*

Fig. 2.4 *The American Wolf* with accessories and painted background; photographed after Hurst's Free Museum opened; *Hurst's Stereoscopic Studies in Natural History, For Schools and Entertainment No. 3*

Stereoviews were used as early as 1857 to depict natural history specimens (Wonders 1993b). When Hurst's high-quality, hand-tinted stereoviews were first produced is unknown. The sheer amount of work involved in mounting the hundreds of taxidermy specimens showcased in his stereoviews is considerable, especially when one recognizes Hurst also recreated, even if temporarily, the natural habitat that frames each specimen grouping. Hurst had a stereoscopic studio that allowed for the construction of temporary habitats, including water. He wrote in 1870: "My interest is to exhibit in every view, as nearly as possible, the locality which every specimen inhabits, whether rocks, woods or waters. All specimens will be colored from the original contained in my private museum." Frederic Webster who lived nearby in Troy, New York, produced a smaller number of stereophotographs starting about 1868. Webster's method is thoroughly described in his autobiography (Webster 1945).

In 1871 Hurst opened the Free Museum to display his extensive collection of birds and beasts. A clipping from an Albany newspaper describes the display (Hatt 1976): "the stuffed toads and animals are surrounded by natural rocks, trees, herbs and foliage, such as distinguish the *habitat* of the animal." It is unclear how many of Hurst's stereoviews depict specific habitat dioramas within the museum and how many represent short-term assemblages made for the purposes of photography. However, we do know that his museum expanded from just 9 Elm Street in Albany to also include 11 and 13 Elm Street. These buildings together would have provided enough room to display a multitude of habitat dioramas in the Free Museum. Regrettably, there are no known photographs of the museum's interior other than the stereoviews.

Hurst produced 60 stereoviews of bird and mammal species of North America. All depict taxidermy arranged within a recreated natural habitat and some include a painted background. Although the sets were promoted for science education in schools, an additional set of 20 stereoviews veered toward a style more consistent with entertainment. Many of the latter were anthropomorphized animals in the style of Ploquet in the Great Exhibition.

By our current definition, these paintings, combined with a taxidermy mount and recreated foreground, fulfill the requirement for a true diorama. Whether they were permanent exhibits is unknown. Multiple versions of the same subject are apparent in the collection of Hurst's stereophotographs. For example, *Hurst's Stereoscopic Studies in Natural History, For Schools and Entertainment No. 3.* portrays an American Wolf collected in 1865 next to a small dog mounted in 1853. This wolf and dog stereophotograph has two different versions. The first dates from before Hurst's Free Museum, the address listed only as 9 Elm Street (Fig. 2.3). A second version was published a few years later when the museum existed and had grown to encompass 9 and 11 Elm Street (Figs. 2.4 and 2.5). Both versions of the display include painted backgrounds but the latter may have been a permanent exhibit. To show the diversity of styles by Hurst, Fig. 2.6 shows six different habitat groups.

Although a habitat diorama vanguard, Hurst was not the only American to attempt habitat groups between 1850 and 1876. William Werner, a taxidermist from Allentown, Pennsylvania, created a series of small bird groupings by replicating, in exact detail, nesting sites. These groupings, begun about 1868, were made for the purpose of display and mounted within cases. They featured a bird's nest, with eggs or young, and the immediate habitat and adults arranged around that centerpiece (Reger 1922). These displays were made in the same style that E. T. Booth had begun about a decade earlier in Brighton, England. Booth is often credited as being the first to introduce groups into a public museum (Lucas 1914a).

In Boulder, Colorado, Martha Maxwell also began building habitat groups around 1868. Maxwell not only built nest groups and trees decorated with multiple

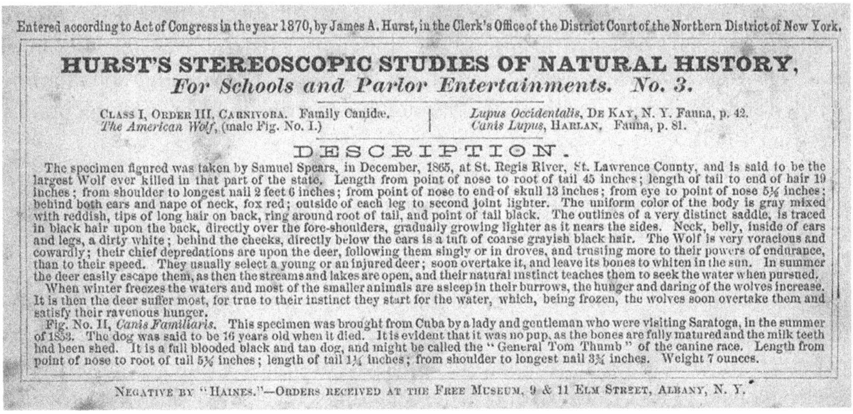

Entered according to Act of Congress in the year 1870, by James A. Hurst, in the Clerk's Office of the District Court of the Northern District of New York.

HURST'S STEREOSCOPIC STUDIES OF NATURAL HISTORY,
For Schools and Parlor Entertainments. No. 3.

CLASS I, ORDER III, CARNIVORA. Family Canidæ.
The American Wolf, (male Fig. No. I.)

Lupus Occidentalis, DE KAY, N. Y. Fauna, p. 42.
Canis Lupus, HARLAN, Fauna, p. 81.

DESCRIPTION.

The specimen figured was taken by Samuel Spears, in December, 1865, at St. Regis River, St. Lawrence County, and is said to be the largest Wolf ever killed in that part of the state. Length from point of nose to root of tail 45 inches; length of tail to end of hair 19 inches; from shoulder to longest nail 2 feet 6 inches; from point of nose to end of skull 13 inches; from eye to point of nose 5½ inches; behind both ears and nape of neck, fox red; outside of each leg to second joint lighter. The uniform color of the body is gray mixed with reddish, tips of long hair on back, ring around root of tail, and point of tail black. The outlines of a very distinct saddle, is traced in black hair upon the back, directly over the fore-shoulders, gradually growing lighter as it nears the sides. Neck, belly, inside of ears and legs, a dirty white; behind the cheeks, directly below the ears is a tuft of coarse grayish black hair. The Wolf is very voracious and cowardly; their chief depredations are upon the deer, following them singly or in droves, and trusting more to their powers of endurance, than to their speed. They usually select a young or an injured deer; soon overtake it, and leave its bones to whiten in the sun. In summer the deer easily escape them, as then the streams and lakes are open, and their natural instinct teaches them to seek the water when pursued.

When winter freezes the waters and most of the smaller animals are asleep in their burrows, the hunger and daring of the wolves increase. It is then the deer suffer most, for true to their instinct they start for the water, which, being frozen, the wolves soon overtake them and satisfy their ravenous hunger.

Fig. No. II, *Canis Familiaris*. This specimen was brought from Cuba by a lady and gentleman who were visiting Saratoga, in the summer of 1853. The dog was said to be 16 years old when it died. It is evident that it was no pup, as the bones are fully matured and the milk teeth had been shed. It is a full blooded black and tan dog, and might be called the "General Tom Thumb" of the canine race. Length from point of nose to root of tail 5½ inches; length of tail 1¼ inches; from shoulder to longest nail 3¾ inches. Weight 7 ounces.

NEGATIVE BY "HAINES."—ORDERS RECEIVED AT THE FREE MUSEUM, 9 & 11 ELM STREET, ALBANY, N. Y.

Fig. 2.5 Rear side of stereophotograph card No. 3 produced by Hurst

species of birds, but also created an elaborate grouping of animals, much like Peale had in 1786. In her instance, Maxwell used species from the western United States and opened her Rocky Mountain Museum to the public in 1874. A writer who visited Maxwell's museum in Boulder published an article in the *New York Independent* in September 1875 (Benson 1986). It described the group but also gave a description of her exhibit techniques:

> She has built up a sort of frame-work, in the shape of rocks. This is covered with a coarse canvas cloth, which has been prepared with glue or some sticky substance. Over this, coarse shining sand of a dark gray color is sprinkled thick: and as the cloth is sticky, the sand remains. At a very little distance nobody would know the rocks from real rocks of gray stone. [Then] she has set real pine and fir trees among them, and little clumps of grasses, and mounds of real dirt (pp 108).

Her concept is not dissimilar to the description of the 1870 American Revised Edition version of Brown's taxidermy manual; only, Maxwell's creation is executed in a much larger scale (Fig. 2.7, left). She corresponded with and acquired reference books from Smithsonian employees Spencer Fullerton Baird and Robert Ridgway and United States government explorers who had visited her museum (Benson 1986). It is likely she either had, or had read a copy of the Smithsonian Revised Edition of Brown's taxidermy manual. In her location and time, large quantities of paper may not have been available from which to create ground-forms. However, canvas, which was readily available for tent construction and other uses, was substituted. Maxwell also sold a number of stereoviews of her Rocky Mountain Museum, much like Hurst and Webster, to help supplement her income; these were dated 1875. An example can be seen in Fig. 2.7, left.

The state of Colorado's centennial commissioners persuaded Maxwell to move her museum exhibit to the Centennial Exposition in Philadelphia in 1876. There she set up an even larger habitat grouping in the Kansas-Colorado Building. This exhibit included a small stream and pond with living fish that were surrounded by taxidermy specimens such as beavers, muskrats, and waterfowl. Above this area were

Fig. 2.6 clockwise from top left: *Brown or Bald Eagle* (Bald Eagle); *Musquash* (Muskrat); *Red Shouldered Buzzard* (Red-shouldered Hawk); *Horned Grebe; Common American Gull* (Herring Gull); *Vir. Partridge* (Northern Bobwhite).

Fig. 2.7 Left—*Maxwell Rocky Mountain Group in Boulder Colorado*; right—*Maxwell "Women's Work" at Philadelphia Centennial Exhibition*

plains and then mountains; according to each life zone in the Rocky Mountains, Maxwell then placed animal mounts such as rabbits, heron, bear, elk, and mountain lion. This exhibit was one of the highlights of the Centennial Exhibit and thousands of stereoview cards were sold to help fund Maxwell's stay there. The stereoviews document her exhibit in detail (Fig. 2.7, right).

The Centennial Exposition had several buildings containing taxidermy aside from the Kansas-Colorado Building. These included a government building that housed displays sent from the Smithsonian. Henry Ward, a successful businessman who dealt in the supply of natural history specimens, was contracted to put exhibits into this building as well. The Agriculture Hall also displayed some very good work from United States taxidermists, including William Werner. George B. Sennett, a prominent ornithologist, summarized the many taxidermy exhibits, including Martha Maxwell's, when he visited the exposition. He considered the work by Werner to be the finest in both science and taxidermy (Deane 1923). Werner "had about twenty-five groups of pairs, young, nest and eggs, each group under glass globes".

George Brown Goode (1895) and later John Rowley (1927a) both reported an exhibit of fur seals, mounted by Smithsonian taxidermists Joseph Palmer and Julius Stoerzer for the 1876 exhibit. The fur seal exhibit was the first large mammal group prepared by Americans for any museum in the country. They both included in their articles an illustration that depicted the seals arranged among recreated rock forms. However, an original stereoview of the exhibit in 1876 shows the seals resting on an unadorned, wooden platform. The artificial rocks

Fig. 2.8 Lion attacking a Dromedary

must have been added later and could have been inspired by other displays at the exposition, such as Maxwell's.

The most famous taxidermy group to be shown in Philadelphia, *Lion Attacking a Dromedary*, (Fig. 2.8) was acquired by the American Museum of Natural History (AMNH) in New York. Representatives of the AMNH traveled in Europe in 1869 to acquire specimens for their new museum opening that year. This group, mounted by Édouard Verreaux (North 1882), was purchased from the Maison Vereaux in Paris. The dramatic display showcases a pair of fierce Barbary Lions as they attack a messenger on camelback in northern Africa. This exhibit stirred the emotions as it had in the earlier Paris Exposition of 1867 where it won a gold medal and was hailed as a major advance in taxidermy display. Considering the numerous well-received taxidermy works, it becomes apparent that the 1876 Centennial Exposition in Philadelphia was a significant milestone in the popularization of mounted habitat groups and sparked an increase in the taxidermy movement in the United States.

2.6 The Society of American Taxidermists

Like many Americans, Henry Ward began a business supplying specimens to collectors and museums beginning around 1860. His business initially specialized in fossils and geological specimens, and Ward traveled throughout Europe acquiring specimens to resell in the United States (Ward 1948). On a visit to the Maison Verreaux, it was suggested that Ward could expand his business by embracing more branches of natural history (Lucas 1914a). The resultant business became Ward's

Natural Science Establishment based in Rochester, New York. As experienced American taxidermists were hard to find in the early 1870s, Ward hired workers from Germany, Italy, and France, as well as young, eager Americans enchanted by natural history (Lucas 1933). One worker from France, Jules Bailley, had spent 15 years working at the Maison Verreaux and was an excellent teacher. Many young Americans who learned their trade at Ward's from the 1870s to the beginning of the twentieth century became museum taxidermists and directors of museums after they left, or "graduated," from Ward's.

It is not within the scope of this chapter to discuss Ward's business, as it is well documented (Ward 1948), but it is important to point out the assemblage of preparators in osteology, taxidermy and other trades who founded a Society of American Taxidermists (SAT). It was conceived in 1880 with this ideal, included in the First Annual Report of the SAT (1881): "The objects of this society are to promote intercourse between those who are interested in the art of Taxidermy in various parts of America and to encourage and promote the development of that art, and to elevate it to a permanent and acknowledged position among the fine arts".

The society produced three annual reports and held important competitions in Rochester, Boston, and New York. The third report lists 95 active members, 12 honorary members, and 3 corresponding members, successfully bringing taxidermy to the forefront in the United States. (SAT 1884).

Two very successful taxidermy groups were exhibited in the first SAT competition. The first president of the society, Frederic Webster, entered a bird group, *The Flamingo at Home* (Fig. 2.9). Webster later claimed that his entry formed the basis

Fig. 2.9 *The Flamingo at home* (Flamingo Group)

Fig. 2.10 *A fight in the tree tops* (Orang Utans Group)

for all subsequent habitat bird groups (Webster 1945). Through previously presented evidence of the work of Peale, Hurst, Werner, and Maxwell, this chapter disputes Webster's assertion while at the same time acknowledging the influence of his creation. William T. Hornaday entered a group of "Orang Utans," titled *A Fight in the Tree Tops* (Fig. 2.10). This grouping had debuted at an American Association for the Advancement of Science meeting in Sarasota Springs, New York, in 1879. A photograph published on the internet shows the completed group as well as Hornaday proudly standing at its side. Hornaday later sold his orangutan group to the Smithsonian Institution and was commissioned by the AMNH to complete a new group of orangutans shortly thereafter. Neither the flamingo nor orangutan group displays had painted backgrounds but both included artificial plants. The Flamingo Group was sold to the Milwaukee Public Museum in 1882 and, while on exhibit, had two different painted backgrounds, up to 1945 when it was still on exhibit (Vance 1946).

However effective these flamingo and or orangutan groups were in influencing the fashion of the time, they are often incorrectly identified as the first taxidermy groups in America. Nevertheless, Webster's and Hornaday's two early SAT habitat groups were accepted into successful museums, and their groupings influenced museums to use habitat accessories and artistic efforts in taxidermy. Only a few years earlier, a Smithsonian researcher had argued against adding habitat materials to otherwise "scientific specimens" (Coues 1874).

2.7 American Advances in Taxidermy Methods and Habitat Accessories

On a visit to England, a trustee from the AMNH convinced two modern practitioners of artificial vegetation to immigrate to the United States (Rowley 1927a). In England, the artists Mrs. E. S. Mogridge and Mr. H. Mintorn, a husband and wife team, had been producing accessories for nesting groups and soon, in collaboration with the taxidermists at the AMNH, began producing elaborate nesting bird cases for the exhibits in New York. The first bird group was an American Robin in 1887 and many followed. Jenness Richardson, an artist who had trained under Hornaday at the Smithsonian, produced the taxidermy for these displays. Hornaday had previously left Ward's in 1882 to begin work as the chief taxidermist at the new United States National Museum (now the National Museum of Natural History, NMNH). Hornaday trained many taxidermists while he worked there.

Although Mogridge and Mintorn were in the United States for only 6 months, modern methods of creating artificial vegetation were passed from the English artists to Jenness Richardson, the chief taxidermist at the AMNH. John Rowley, then working at the AMNH under Richardson, also learned these techniques and later took over as head taxidermist at the AMNH after his mentor died in 1893.

While the AMNH had fully embraced the use of artistic accessories in taxidermy, which would later lead to their dominance of habitat diorama work primarily initiated by Frank Chapman and continued by Carl Akeley, the NMNH did not entirely agree with the fashion. George Brown Goode (1895), in assessing the level of American vs. English taxidermy and exhibit styles in 1893, had this to say:

> We frankly admit that in the matter of environmental groups of birds, Great Britain still surpasses the United States. So far as taxidermy is concerned, American workmen can hold their own, but the art of making and grouping accessories we have yet to acquire. The only successful accessory work done in this country is that by the Mogridges, who were trained at South Kensington, and who are represented extensively in the New York Museum and one piece in Washington.
>
> Many of the groups of this kind, even when made by the Mogridges, error in making the accessories more prominent than the birds and filling the cases with artificial flowers and leaves to such a degree that the birds are entirely subordinate…. (pp 46).

The NMNH and the AMNH were the largest museums in the United States at that time. In the 1890s, taxidermists at each institution produced taxidermy manuals. William T. Hornaday, of the NMNH, published *Taxidermy and Zoological Collecting* that totaled 362 pages (Hornaday 1891). This monumental work advanced the methods of taxidermy like no previous publication had in America. One chapter was devoted to ornamental taxidermy and in it, only part of one page described artificial leaves; this half-page primarily reveals where they could be procured. Hornaday did have a good section on groups and covered the recent history of the SAT and the development of exhibit displays. John Rowley of the AMNH published the *Art of Taxidermy* in 1898 (Rowley 1898). This taxidermy text of 244 pages included the first detailed information on crafting vegetation published in the United States. Chapter 9 titled "The reproduction of foliage, etc., for groups" has 38 pages of text on the manufacture of wax and paper leaves and flowers with numerous illustrations and two photographic plates. The dichotomy in these publications leads us to conclude that the AMNH valued both dioramas and science, whereas the NMNH concentrated more exclusively on science.

With these two publications, the large natural history museums that started after 1890 had the information available from the outset to build integrated displays of groups of animals within habitats. They readily embraced this maturing exhibit style. Many newly establishing natural history museums built elaborate exhibits with numerous, accessorized habitat groups in their first decades, including The Field Museum of Natural History (FMNH) in 1893, the Carnegie Museum of Natural History (CM) in 1895, the Denver Museum of Natural History in 1900, and the Los Angeles County Museum of Natural History in 1913. Large museums that had started earlier also shifted to this style of exhibit, most notably the Academy of Natural Sciences of Philadelphia in 1812, California Academy of Sciences in 1868, and the Milwaukee Public Museum in 1882.

The era following the advancements made by the SAT and the development of habitat diorama artistry at multiple United States museums has been well documented in papers specified earlier and in a recent thesis and dissertations (Andrei 2006; Cain 2007; Smith 2012; Jones 2014). By 1900, accessorized habitat groups with painted backgrounds became the emerging fashion, advancing the displays to what we currently define as true habitat dioramas. The quality, quantity, and breadth of habitat dioramas became the standard by which a great natural history museum was measured.

2.8 Carnegie Museum of Natural History Dioramas Through the Years

Early development of habitat dioramas at CM followed a path that was typical for this emerging group of institutions. Although CM was dedicated in November 1895, the museum itself did not open any exhibits until a year later and most of these were gifts or loans. Some of the latter were on loan from Ward's. In spring 1897, CM hired Frederic Webster to take charge of the Department of Preparation. In his years before the CM, Webster had been employed at Ward's and worked extensively with Hornaday. He followed Hornaday to Washington, D.C., in the early 1880s, and around 1890 moved to New York where he opened a studio and supply business. Webster brought many specimens from his supply business to Pittsburgh when he was hired and sold them to the CM. A group of orangutans were purchased from Ward's but these were far inferior to those produced by Hornaday two decades earlier.

Fig. 2.11 Condor group

By 1900, Carnegie Museum's exhibits were flourishing with newly created habitat groups fabricated by Webster and his assistant Gus Link. They began building the same style of accessorized habitat groups that had elicited excitement at the SAT competition, AMNH, and NMNH. A group of five flamingoes and a series of bird nesting groups filled out the gallery of birds. A large, four-sided, glass display case gathered rave reviews for its dramatic depiction of a recumbent elk apparently killed by a Native American arrow; Condors and Turkey Vultures surround and investigate the corpse. (Fig. 2.11). Hornaday describes the exhibit as a masterpiece in his review article (Hornaday 1925). CM artist Ottmar F. von Fuehrer added a background painting to this case in the 1930s. In the gallery of mammals, a group of Fur Seals were assembled similar to that in the NMNH illustrated by Goode in 1895. In a major coup of sorts, CM was able to rescue the Arab Courier exhibit from the AMNH, where it had fallen into some disrepair, and moved it to Pittsburgh. Webster and Link restored it to its former glory when it was shown both in Paris in 1867 and at the centennial exhibit in Philadelphia. The grouping is still a favorite among many visitors to CM. (Fig. 2.8). Webster's tenure at CM lasted only 10 years.

Shortly after Webster's time, two talented taxidermists, both with experience from Ward's, joined the preparatory department at the CM. Brothers Remi and Joseph Santens were among the best taxidermists in the nation at the time. Remi had been the head of the taxidermy department at Ward's for many years. In the dozen

Fig. 2.12 Savannah watering hole

years both brothers worked at CM, they built a large collection of four-sided cases that displayed some technically and artistically adept examples of taxidermy. Many were highlighted in the review article by Hornaday in 1922. One of those, a four-sided Jaguar case, is still shown in the manner in which it was displayed in 1910, the only differences being new glass and a lighting hood. The four-sided cases built between 1912 and 1920 housing African game were exhibited as such until a complete diorama renovation of African Hall was opened in 1993 primarily through the efforts of Chief Exhibits Preparator Pat Martin (Fig. 2.12).

Carnegie Museum of Natural History began building classic dioramas with curved ceilings and backs in 1923, the first being a Horned Owl and Skunk case featuring a painted background by George Miksch Sutton (Sutton 1980). Sutton was a respected ornithologist and noted artist. The museum is interested in reinvigorating its historic dioramas as well as the modern engagement potential they represent and recently saved this diorama and six others built between 1923 and 1933 and built a new exhibit area entitled *The Art of the Diorama.* To create a more immersive experience the small gallery is fashioned to suggest a Victorian museum, complete with chandelier and mahogany furniture. Included are two bubble-glass wall cases that are uniquely American mini-dioramas, once popular as natural history home décor. Unlike many modern dioramas, which use ground forms and vegetation to blend the barriers between case and visitor immersion, *The Art of the Diorama* focuses on the visitors of a century ago, blurring the boundary between cultural and natural heritage.

The exhibit also includes a physical interactive in which families can assemble their own dioramas using child-friendly materials made available to them. In the exhibit, Anna Dierdorf and Hannah von Fuehrer, who preserved or created vegetation for the CM dioramas, were—after many decades—recognized for their contributions.

2.9 Contemporary Status of Habitat Dioramas in Museums in the United States

In recent times, museums have come to question the continued maintenance and display of their historic dioramas. During the late 1990s and 2000s, dioramas experienced a decline in popularity in the United States and elsewhere and many historic dioramas were disassembled to make room for more contemporary methods of gallery storytelling. Often the accessioned taxidermy remains in displays and collections, but many of the groups, accessories, and painted backgrounds have been destroyed. This low point in the perception of dioramas paralleled an escalation in the multimedia-savvy industry of contract museum consultants. Today, many historic displays may be threatened by architectural renovations or exhibit designers and architects unfamiliar with the artistry, historical significance, or cultural resonance of natural history dioramas. The knowledge base to create new ones has moved outside of museums and very few artists are being trained in the traditional museum methods.

To build a new diorama from the ground up would require such uncommonly specialized technicians that the cost might be prohibitive. As exhibit departments shrink in the modern, streamlined museum and contractors are brought in to redesign entire galleries and even whole museums, historic dioramas need advocates to protect them from being dismantled. Because of these conditions, CM considered it worthwhile to conduct an informal survey of many of the larger museums in the United States to assess the current status of habitat diorama exhibits. Some museums have pulled apart large numbers of dioramas in the last 20 years, some have displays under threat of destruction, and a few museums are building new exhibits in this style. A survey was sent out to a number of United States museums that house dioramas in an exercise to ascertain how dioramas currently fair in museums. In the paragraphs below we summarize surveys from the Oakland Museum of California, the Natural History Museum of Los Angeles County, the Yale Peabody Museum of Natural History, and the Denver Museum of Nature and Science. We also interpret research on the California Academy of Science, the American Museum of Natural History, the National Museum of Natural History, the Field Museum of Natural History, and the Bell Museum of Natural History.

In 2008, The California Academy of Sciences (CAS) moved from its original location to a new building. During reconstruction, only 12 of the 24 original dioramas in the Simpson Africa Hall (Anon. 1944) were transferred and rebuilt in the new Renzo Piano building. The rear side of the Africa Hall is now devoted to a large enclosure for live penguins and other live animals have been introduced around the displays. All 20 original dioramas in the North American Hall (Anon. 1939) were dismantled. The exhibition design work was accomplished by outside firms rather than an institutional team of personnel. The *Tushan Africa Center* exhibition and its integration were reviewed in the National Association for Museum Exhibition's peer-reviewed journal, *Exhibitionist* (Schwarzer et al. 2009) and described as "confusing, if unsettling" but also "when combined with live animals…it will

Fig. 2.13 Pronghorn diorama at OMCA. (Credit Ramon Felix)

become a favorite gallery for a new generation of children." It seemed a real challenge to update the interpretive experience of the gallery, to maintain the cultural heritage of the dioramas, and to also integrate live animals all at the same time.

The Oakland Museum of California (OMCA) was able to make use of some specimens from two cases in the CAS's dismantled North American Hall. Alicia Goode, one of the few remaining museum staff taxidermists, and other workers at OMCA spent almost 5 years working on the new Gallery of California Natural Sciences that reopened in 2013. That new gallery used and refurbished most of the original dioramas from the former gallery. Two new dioramas incorporated three of the original five Tule Elk mounted by John Rowley and a pair of Pronghorn Antelope. Highly skilled restorations were done to fill patches of skin that were missing due to prior pest damage and to re-color the fur on the new Tule Elk diorama. The lava tube diorama features two previous CAS antelope, a mixture of pre-existing taxidermy from OMCA, and some new models and taxidermy. To flesh out the diorama, a trip with the recently retired curator was made to the Lava Beds National Monument research station to understand the area's geological formations and landscapes. At the site, researchers made rock molds, collected plant material and invertebrates, detailed field notes and species lists, and created preliminary drawings and *plein air* paintings. Three people were involved in concept, design, and fabrication of the diorama: Alicia Goode, rock sculptor Jenny Cole, and mural artist Cleo Vilett. Except for some contracted interior plasterwork, the diorama was made entirely in-house and features a walk-through lava-tube cave that incorporates a viewing window into the pronghorn diorama (Fig. 2.13).

Like OMCA, the Natural History Museum of Los Angeles County (NHMLAC) has had an active diorama program continuously since the 1920s. Currently three of the four classic natural history diorama halls are open, which, together with a bird hall that opened in 1991, has 69 dioramas on view. Taxidermist Tim Bovard is currently the lead in diorama design, fabrication, and installation for NHMLAC and uses a number of volunteers in his work. Bovard's position as a diorama artist was once common in most exhibition departments but now he is one of the last in his profession employed by a museum. When he began work in the 1980s, there was a larger habitats team including James D. Olson and Charles Fischer, who excelled in design and fabrication, and the background artist Robert R. Reid, who retired in 2014.

Many of the dioramas at NHMLAC have been updated in the last 10 years. Between 2005 and 2006, most of the first-floor dioramas in the African and North American halls were renovated, adding about 80 new mounts, railings, diorama labels, and lighting. The second floor of North American Hall was completely refurbished, opening in 2007, with newly painted backgrounds and foregrounds and 4 new dioramas. Except for some basic electrical work and carpentry, all of the work was accomplished in-house. NHMLAC used a number of modern innovations including animatronics, sound and lighting effects, projections, and touchable materials, to help tell the story while capturing visitor interest in the diorama halls.

The American Museum of Natural History (AMNH) is the largest natural history museum in the United States and has significant holdings of many of the world's finest historic habitat dioramas. They value them immensely. Even this large, well-endowed museum has not had a staff taxidermist since 1988. Rather, they contract with various taxidermists to create mounts but have not commissioned a full-sized zoological diorama in decades. The cost of producing new dioramas is significant, almost prohibitive, as it was during the heyday of diorama work in the United States when the preparators who built them poured their hearts, as well as years of work, into each of their creations (Cain 2007). Because they value their dioramas and treat them as works of art, the AMNH has led the way in developing new methods to refurbish them. They have used social media and scientific blogs to focus attention on their work. The AMNH created a series of 16 short videos explaining various steps in restoring the Hall of North American Mammals. They are wonderful resources and are widely available on YouTube (AMNH.org 2012).

During the AMNH's process of restoring the Hall of North American Mammal's 43 dioramas in 2012, a $2.5 million project, they researched the best methods for re-coloring fur, retexturing nose pads, and improving the displays in the 70 year-old hall. They published some of this research together with Yale University and currently have a 3-year IMLS grant titled "Recoloring Faded Taxidermy: Research into the Properties and Applicability of Dye Materials for Conservation Treatment." Since 2014 they have posted continual updates to their improving techniques in an online blog (AMNHCONSERVATOR 2016). We imagine that the attention the AMNH has given to its North American Mammal Hall, and the research they have conducted in their process, may influence diorama renovations at AMNH and at other US natural history museums in the future.

The National Museum of Natural History (NMNH) demonstrated a very different attitude toward their habitat dioramas when they rebuilt their mammal hall in 2003.

To create the Kenneth E. Behring Family Hall of Mammals, they dismantled their historic dioramas, many that approached 100 years of age, and installed what they considered a "more sophisticated" and family friendly presentation of the content (Brennan 2008). Two talented taxidermists, John Matthews and Ken Walker, were hired for the project, who, along with their then staff taxidermist Paul Rhymer, and occasional help by others, built a mammal hall with very little habitat accouterments. Each taxidermy specimen is placed in only minimal context with its environment. The drama of the display relies on the position and gesture of each piece of lone taxidermy. Although a few have child partners such as their manatee, or prey partners, such as their leopard and impala, the majority of the mammals are represented singly and have a painted beige backdrop and labels. It is a minimalist, architectural presentation of the specimens. Paul Rhymer left the museum a few years later and the position was not replaced. A review of the exhibition can be seen in the Spring, 2008 AAM journal *Exhibition* (Brennan 2008). The review is mainly positive although it does note that "many fans of the old dioramas were sorry to see them go" and "more than half the visitors interviewed…[identified the historic dioramas] as their favorite thing in the museum".

Historic dioramas, like those in the original NMNH mammal hall, were not built to be moved. In fact, they are often designed into the interior architecture of their museums. Dioramas can be very large and, when they include domed plaster backdrops, weigh thousands of pounds. They are fragile and difficult to move without damaging or destroying them. Dioramas are most at risk when a natural history museum is renovating its architecture or moving into a new building. That risk was true of the CAS, described above, and is also true of the Bell Museum of Natural History.

In 2014 The Bell Museum at the University of Minnesota developed plans for a new museum building to replace the old building specifically built to house its' dioramas. There was speculation that the dioramas might not move to the new location (Quinn 2014). That same year, a "Save the Bell Museum" Facebook page was created and a series of GigaPan images were published online to show the outstanding details and quality of the many dioramas at the Bell (Nelson 2014). Their plans were to document the conservation process and to share what they learn during their move with other natural history museums. In July 2018, the museum completed the move of many of the exquisite dioramas to the new building and published a blog regarding the move (Bell Museum: Windows to Nature, 2018).

Diorama advocates are using social media technologies to help spread their enthusiasm for the medium. Most museums have Facebook, Twitter and Instagram accounts, and many also produce YouTube videos. In the last 3 years, the Field Museum of Natural History in Chicago has used its social media campaign to the fullest. Emily Graslie, who had created a YouTube series of educational videos called The Brain Scoop, was hired in 2013 as the chief curiosity correspondent of the FMNH. To this date, she has created over 130 videos and has 32,800 followers on her Twitter account. Her "Brain Scoop" YouTube channel has 476,640 subscribers, and has had 24,616,120 views as of August 2018. In April of 2015, she initiated an IndieGoGo project to raise money to build a diorama, which had a space designated in a hall where 19 of 20 dioramas had been completed decades earlier. Four hyenas, mounted by Carl Akeley after his 1896 trip to Somaliland, existed in a four-sided mahogany case in Reptile Hall. She used a Brain Scoop video to steer potential funders to the URL (Graslie 2015). By the end of funding, Graslie and the FMNH were able to raise $155,165 with the help of over 1500 contributors, which was used to build the first diorama at the FMNH in 60 years. This project was discussed on a National Public Radio show on dioramas featuring Graslie and Michael Anderson, the preparator from the Yale Peabody Museum of Natural History (YPM) (McNicol 2015).

Like Graslie, Anderson also created a recent diorama all the while documenting his process and promoting it through social media. He has been at the YPM for 28 years following a 5-year stint at the AMNH. Anderson persuaded the Canadian Museum of Nature, which had a fiberglass diorama shell painted by James Perry Wilson, who is often regarded as the finest diorama background painter of all time (Anderson 2015), to donate the unused shell to the YPM. Anderson used this shell in constructing the Point Pelee diorama, and created a blog in December 2009 about its building, continuing to update it until the diorama's opening in very early 2011, detailing every step of the process (Anderson 2009). Anderson publishes emerging techniques of model production on a separate blog (Anderson 2010). In it, for example, Anderson details his new method of making leaves using photographic prints of leaves on fabric embedded with clear epoxy. By eliminating the need to paint them, which destroys the translucency of the leaves, Anderson has created a more realistic example of natural foliage. In 2014, Anderson was relieved of day-to-day exhibit responsibilities and took on the task of restoring YPM's 12 habitat dioramas and the Connecticut Bird Hall.

One museum that has particularly embraced their historic dioramas is the Denver Museum of Nature and Science (DMNS). Not only has the DMNS retained most of their dioramas, they have also augmented the presentation to meet and exceed contemporary museum standards for visitor experience. In 2011, the DMNS updated interpretation and gallery environmental effects in one wing of their North American Wildlife Halls. The dioramas were cleaned and lighting was improved, but the bulk of the effort focused on improving the visitor experience. The DMNS revised their labels, making them less wordy and less academic and included a number of original illustrations to help guests pick out more diorama details. Additionally, a few videos, several environmental sound effects, and several models or touchable specimens were added to increase types of access to the gallery. While cleaning and repairing their open, Prehistoric Journey dioramas in 2014, the DMNS was at the same time updating interpretation in their Bears and Sea Mammals Wildlife Hall.

Very little work other than some lighting adjustments was done within the dioramas proper, but all new interpretation replaced outdated labels. Additionally, several exhibit cases, touch specimens, and push-button activated sound effects were added to increase immersion and types of access to the information. Faux-rocks and a faux-tree, featuring respectively a bronze sea lion pup and bronze bear cubs were added for immersion, photo-op, and touch-experience options. Additionally, a digital, vertically-projected stream bed was added to the center floor of the exhibition.

Situated between seating shaped like rocks, guests could attempt to successfully "catch" a salmon in the stream bed, similar to the bears featured in the nearby dioramas (Schoemer pers. comm. 2016).

The DMNS is known for the high quality of its historic dioramas; the background painting, taxidermy, and dramatic groupings are particularly handsome. Perhaps it is pride in the quality of its holdings that helped influence DMNS's museum culture to augment, improve, and modernize dioramas rather than remove or destroy them in favor of different styles of exhibition. As wildlife dioramas become less central to natural history interpretation, and also less common, the attention to these holdings may differentiate them in times ahead.

The future of the habitat diorama in United States natural history museums is in question. Although many museums still have groupings and dioramas, whether through choice or inattention, our surveys show that new dioramas are not being planned. The majority of United States museums no longer have the talent in-house to build new dioramas. Those contractors or taxidermy studios still practicing the art form are being hired by private collectors and by for-profit companies. For example, Cabella's, a hunting and fishing outfitter in the United States and Canada, has commissioned elaborate taxidermy groupings and walk-through wildlife experiences for the patrons of their many stores. New dioramas may be cost prohibitive to museums, but a version of the skill-set is still available should the interest and funds support new diorama initiatives. If taxidermy and the collection of display animals fall outside of a museum's mission due to ethical concerns, there are contractors who create beautiful, life-like reproductions.

Although plans for new dioramas are uncommon, augmenting and improving existing dioramas has proven to be quite successful. Improvements to lighting greatly refresh a tired diorama. Visitation to the CM Hall of Botany greatly increased with a minimal investment in new LED lights within the dioramas and by adding modern carpet tile in the gallery. Improvements mean that color rendering in the dioramas looks more true to life, making the gallery feels brighter, more modern, and, arguably, more relevant as a result of the changes.

Classic diorama presentation can be a repetitive experience for visitors but this can be improved by offering a variety of visitor experiences within the hall. In partnership with the University of Pittsburgh, CM has conducted visitor testing in our wildlife halls. From observation, visitors race through the gallery, glancing at each diorama, spending a minimal amount of time with them. Almost any intervention, 8 were tested, drives visitors to stay longer at dioramas and some interventions cause visitors to look more closely (Knutson et al. 2016). In particular, asking visitors to classify species within a diorama holds their attention, but most activities out of synch with the rest of the presentation grabbed our visitors' attention.

Each natural history museum is unique. It would be a mistake for all museums to fill the entirety of their galleries with habitat dioramas. Just as a symphony has many movements that differ in style, mood, or tempo, museums need to provide a varied experience to keep visitors fresh and engaged. A museum with nothing but dioramas would suffer from a lack of expressiveness, but dioramas are a wonderful note to be able to play. Dioramas should be a valued part of natural history presentation. They connect us to our cultural heritage as institutions and demonstrate how perspectives change over time. Some dioramas can be recognized today as artistic masterpieces and should be accessioned and maintained for posterity. It is unlikely, in the short term, that museums will be able to afford entire departments devoted to the creation of dioramas like they once did in the first half of the twentieth century. However, it is possible that fashions will change and the specialized work will return to museum exhibit departments. In the meantime, museums should look closely at their dioramas and consider them carefully when planning new exhibits, entire exhibition galleries, or museum buildings. Although threatened, it is likely habitat dioramas will continue as a strength of natural history presentation in United States museums.

Acknowledgements We thank Michael Anderson, Tim Bovard, Alicia Goode, John Janelli, and Jody Schoemer for their insight and encouragement; Linda Schreiber, William Schreiber, and Mary Ann Schmidt, ELS for help with the manuscript. We also thank the Hunt Institute for Botanical Documentation, Carnegie Mellon University, for scanning images.

References

Allen, G. (1885). How to collect, mount and preserve all varieties of beast, birds, insects, etc. etc. In *The taxidermist manual*. New York: AL Burt.

AMNHCONSERVATOR. (2016). *In their true colors*. AMNH, 24 April 2016. https://intheirtruecolors.wordpress.com/. Accessed 02 May 2016.

AMNHorg. (2012, September 21). *Restoring dioramas in hall of North American mammals*. YouTube. https://www.youtube.com/watch?v=4LxTKdFaeS0. Accessed 02 May 2016.

Anderson, M. (2009). James Perry Wilson. *James Perry Wilson*. Michael Anderson, 05 Nov. 2009. https://jamesperrywilson.wordpress.com/. Accessed 02 May 2016.

Anderson, M. (2010) Museum model making at Yale Peabody. *Museum Model Making at Yale Peabody*. N.p. https://museummodelmaking.wordpress.com/. Accessed 02 May 2016.

Anderson, M. (2015). James Perry Wilson: Shifting paradigms of natural history diorama background painting. In S. D. Tunnicliffe & A. Scheersoi (Eds.), *Natural history dioramas: History, construction and educational role* (pp. 67–80). Dordrecht: Springer.

Andrei, M. A. (2006). *Nature's mirror: How the taxidermists of Ward's Natural Science Establishment transformed wildlife display in American natural history museums and fought to save endangered species*. Dissertation, University of Minnesota.

Anon. (1853). *Official catalogue of the New York exhibition of all industry of all nations*. New York: George P. Putnam & Co.

Anon. (1887). *Ladies manual of art or profit and pastime. A self teacher in all branches of decorative art, embracing every variety of painting and drawing on china, glass, velvet, canvas, paper and wood. The secret of all glass transparencies, sketching from nature, pastel crayon drawing, taxidermy, etc*. Philadelphia: American Mutual Library Association.

Anon. (1890). *The taxidermist's guide. A manual of instruction in the art of collecting, preparing, mounting and preserving all kinds of animals, birds, fishes, reptiles and insects*. Adapted for the use of amateur travellers and practical workers. New York: Hurst & Co.

Anon. (1939). *The North American Hall of the California Academy of Sciences*. Dedicated September 22, 1916. San Francisco: California Academy of Sciences.

Anon. (1944). *The Simpson African Hall of the California Academy of Sciences*. Dedicated December 14, 1934 (2nd ed). San Francisco: California Academy of Sciences.

Anon. (n.d.-a) In Bowdich, S.W. (1820). *Taxidermy: or, the art of collecting, preparing, and mounting objects of natural history, for the use of museums and travelers* (1st ed). London: Longman, Hurst, Rees, Orme, and Brown.

Anon. (n.d.-b). In Brown, T. (1870). *The taxidermist's manual; a complete practical guide to collecting, preparing, and preserving all kinds of animals, birds, reptiles, insects, etc.* Designed for both experts and amateurs. Smithsonian Institution. America revised edn. From twentieth English. New York: The Orange Judd Company.

Batty, J. H. (1880). *Practical taxidermy and home decoration; together with general information for sportsmen* (reprinted ed). New York: Orange Judd Co.

Beaver, D. (1999). Writing natural history for survival—1820–1856: The case of Sarah Bowdich, later Sarah Lee. *Archives of Natural History, 26*(1), 19–31.

Benson, M. (1986). *Martha Maxwell, Rocky Mountain naturalist*. Lincoln: University of Nebraska Press.

Boitard, P. (1825). *Manuel du naturaliste préparateur ou l'art d'empailler les animaux et de conserver les végétaux et les minéraux*. Paris: Roret.

Brennan, B., Singer, R., & Pekarik, A. (2008, Spring). The Kenneth E. Behrin family hall of mammals. *Exhibition,* 87–94.

Brown, T. (1833). *The taxidermist's manual; or the art of collecting, preparing, and preserving objects of natural history. Designed for the use of travellers, conservators of museums and private collectors*. Glasgow.

Cain, V. E. M. (2007). *Nature under glass: Popular science, professional illusion and the transformation of American Natural History Museums, 1870–1940*. Dissertation, Columbia University.

Coues, E. (1874). *Field ornithology. Comprising a manual of instruction for procuring, preparing and preserving birds and a checklist of North American birds* (1st ed.). Salem: Naturalists' Agency.

Davidson, H. E. (1885). *Davidson's ichthytaxidermy. An improved method of preserving fishes*. New York: International Bank Note Co.

Davie, O. (1882). *The naturalist's manual, containing descriptions of the nests and eggs of North American birds (Turdidae Tanagridae). Also, instructions for collecting and preserving birds, nests, eggs and insects*. Columbus: The Columbus Printing Works.

Davie, O. (1894). *Methods in the art of taxidermy* (Subscription ed). Columbus: Author.

Deane, R. (1923). Extracts from the field notes of George B Sennett. *Auk, 40*(4), 626–633.

Dufresne, L. (1820). *Taxidermie, ou, L'art de préparer et de conserver la dépouille de tous les animaux, pour les musées, les cabinets d'histoire naturelle, etc*. Paris: Chez Deterville.

Farber, P. L. (1977). The development of taxidermy and the history of ornithology. *Isis, 68*(244), 550–566.

Goode, G. B. (1895). *Report upon the condition and progress of the U.S. National Museum during the year ending June 30, 1893*. In Report of the U.S. National Museum for 1893 (pp. 1–334).

Graslie, E. curator. (2015). Project Hyena Diorama. In *Indiegogo*. Field Museum of Natural History. Retrieved 02 May 2016, from: http://www.indiegogo.com/projects/project-hyena-diorama

Harding, S. (1875). *The amateur trapper and trap makers guide. A complete and carefully prepared treatise on the art of trapping, snaring, and netting*. Containing plain directions for constructing the most approved traps, snares, nets, and dead-falls; the best methods of applying them to their purposes; and the most successful baits for attracting all kinds of animals, birds, &c., with their special uses in each case; introducing, also, practical receipts for preparing them for future use; with concise but comprehensive instructions for preserving and stuffing specimens of birds and animals in the most durable manner. The entire work is based on the experience of the most successful trappers, and on information derived from other authentic sources. New York: Dick & Fitzgerald.

Hatt, R. T. (1976). James A. Hurst, New York's first state taxidermist. New York State Museum. *Journal NAHO, 9*(1), 3–17.

Holder, R. H. (1861). Taxidermy. Directions for collecting and preserving specimens in ornithology. *Transactions of the Illinois Natural History Society, 1*, 597–603.

Hornaday, W. T. (1891). *Taxidermy and zoological collecting. A complete handbook for the amateur taxidermist, collector, osteologist, museum builder, sportsman, and traveller*. With chapters on collecting and preserving insects by William J. Holland (1st ed). New York: Charles Scribner's Sons.

Hornaday, W. T. (1922). Masterpieces of American taxidermy. *Scribner's Magazine, 72*(1), 3–17.

Hornaday, W. T. (1925). Masterpieces of American bird taxidermy. *Scribner's Magazine, 78*(3), 261–273.

Jones, E. B. (2014). *Monarch of the plains: Federalism and ecology in nineteenth century American museum habitat groups*. Thesis, University of Florida.

Kettell, S. (1831). *Manual of the practical naturalist; or directions for collecting, preparing, and preserving subjects of natural history, containing instructions and recipes according to the most approved methods for taking and stuffing quadrupeds, birds, fishes, reptiles*. Boston: Lilly and Wait, and Carter, Hendee & Babcock.

Kingsley, J. S. (1882). *The naturalist's assistant. A handbook for the collector and student with a bibliography of fifteen hundred works necessary for the systematic zoologist*. Boston: SE Cassino.

Knutson, K., et al. (2016). Flexible interventions to increase family engagement at Natural History museum dioramas. *Curator: The Museum Journal, 59*(4), 339–352.

Lucas, F. A. (1914a). The story of museum groups. Part 1. *The American Museum Journal, 14*(1), 1–15.

Lucas, F. A. (1914b). The story of museum groups. Part 2. *The American Museum Journal, 14*(2), 50–65.

Lucas, F.A. (1933). *Fifty years of museum work*. Autobiography, unpublished papers and bibliography. New York: American Museum of Natural History.

Manton, W. P. (1882). *Taxidermy without a teacher; comprising a complete manual of instructions for preparing and preserving birds, animals and fishes with a chapter on hunting and hygiene; instructions for preserving eggs and making skeletons and a number of valuable receipts* (2nd ed.). Boston: Lee and Shepard Publishers.

Maynard, C. J. (1873[1870]). *The naturalist's guide in collecting and preserving objects of natural history, with a complete catalogue of the birds of eastern Massachusetts*. Boston: Fields, Osgood & Co.

Maynard, C. J. (1883). *Manual of taxidermy. A complete guide in collecting and preserving birds and mammals*. Boston: SE Cassino and Co.

McConnaughay, C. J. (1898). *Barbering and taxidermy. (Part first: Barbers' manual. Part second: Text book on taxidermy)*. Chicago: Author.

McNicol, J. (2015). *An up-close look behind the glass of... dioramas*. WNPR News. Colin McEnroe. http://wnpr.org/post/close-look-behind-glass-dioramas#stream/0. Accessed 02 May 2016.

Nelson, T. (2014). *Gigapans*. Panoramic Photography. Gigapan. http://www.gigapan.com/gigapans?order=most_popular&tags=bell%2Bmuseum. Accessed 02 May 2016.

North, F. H. (1882). The taxidermal art. In *The Century Illustrated Monthly Magazine*, December 1882, Vol. XXV(2) (pp. 230–239). Union Square, New York: The Century Co.

Quinn, S. (2014, June 20). Save the Bell Museum. *Save the Bell Museum*. [Facebook status update]. Retrieved from https://www.facebook.com/belldioramas

Reed, C. K. (1900). *Guide to taxidermy with full instructions how to prepare and mount birds, animals and fish. Also a complete list of all North American birds, with prices of their eggs, skins and mounted specimens, also an exhaustive line of ornithologists', oologists' and taxidermists' supplies; valuable information for the amateur, recipes, etc*. Worcester: Charles K. Reed, Taxidermist.

Reger, H. S. (1922). An appreciation of William H. Werner. *Oologist, 39* (suppl 426).

Rogers, S. P. et al. (1989). *An annotated bibliography on preparation, taxidermy, and collection management of vertebrates with emphasis on birds*. Pittsburgh: Carnegie Museum of Natural History, Special Publication 15.

Rowley, J. (1898). *The art of taxidermy*. New York: D. Appleton and Co.

Rowley, J. (1927a). The development of museum taxidermy, part 1. *Mus Graphic (Los Angeles Mus), 1*(4), 121–134.
Rowley, J. (1927b). The development of museum taxidermy, part 2. *Mus Graphic (Los Angeles Mus), 1*(5), 193–206.
Schwarzer, M., Margaret, K., & Pressley, S. (2009, Spring). The Tushan Africa Center at the California Academy of Sciences. *Exhibition*, 75–83.
Sellers, C. C. (1980). *Mr. Peale's museum. Charles Willson Peale and the first popular museum of natural science and art*. New York: WW Norton & Company Inc.
Shufeldt, R. W. (1894). *Scientific taxidermy for museums (based on a study of the United States government collections)*. In Report of the U.S. National Museum for 1892 (pp. 369–436). Washington, DC: U.S. Government Printing Office.
Smith, D. (2012). *Animals and artifacts: Specimens exchanges and displays in Yellowstone National Park, the National Museum, and the National Zoo, 1846 to 1916*. Dissertation, Montana State University.
Society of American Taxidermists. (1881). *First annual report American Taxidermists 1880–81*. Rochester: Daily Democrat and Chronicle Book and Job Print.
Society of American Taxidermists (Hornaday, WT, comp). (1882). *Second annual report American Taxidermists 1881–82*. Judson J. Rochester, New York: Whitel, Book and Job Printer.
Society of American Taxidermists (Lucas, FA, comp). (1884). *Third annual report of the Society of American Taxidermists, March 24th, 1882, to July 1, 1883*. Washington, DC: Gibson Bros. Printers.
Sutton, G. M. (1980). *Bird student, an autobiography*. Austin: University of Texas Press.
Sylvester, S. H. (1865). *The taxidermists' manual, giving full instructions in mounting and preserving birds, mammals, insects, fishes, reptiles, skeletons, eggs, &c* (3rd ed.). Middleboro: The author.
Vance, A. (1946). From the table with fringe on top. *Frontiers, 10*(3), 66–69.
Walker, C. A. (1870). Hints on taxidermy. *Amer Naturalist*, 3, 136–146, 189–201, 481–485.
Ward, H. A. (1881–86). *Ward's Natural Science Bulletin*, 1(1),16 pp, 1(2),16 pp, 1(3),16 pp, 1(4),16 pp, 2(1),16 pp, 2(2),16 pp, 3(1),16 pp, 3(2),16 pp, one unnumbered at 16 pp. Rochester, New York: Ward's Natural Science Establishment.
Ward, R. (1948). *Henry A. Ward, museum builder to America*. Rochester: The Rochester Historical Society Publication Fund Series.
Warren, B. H. (1896). *Taxidermy. How to collect, skin, preserve and mount birds*. The game and fish laws of the Commonwealth of Pennsylvania, Bulletin No. 6. Department of Agriculture, Division of Economic Zoology, Harrisburg, Pennsylvania.
Webster, F. B. (1885–1886). *Practical taxidermy*. Ornithologist and Oologist, 10, 137–139, 153–155, 169, 170, 186–189; 11,3–6, 26–28, 41–43, 57–59, 73, 74, 89, 105–107, 122, 137, 169, 170.
Webster, F. S. (1945). The birth of habitat bird groups, reminiscences written in his ninety-fifth year. *Annals of Carnegie Museum, 30*, 97–118.
Wonders, K. E. (1993a). Bird taxidermy and the origin of the habitat diorama. In R. G. Mazzolini (Ed.), *Non-verbal communications in science prior to 1900* (pp. 411–447). Firenze: Leo S. Olscki.
Wonders, K. E. (1993b). *Habitat dioramas. Illusions of wilderness in museums of natural history*. Acta Universitatis Upsaliensis, Fig. Nova Series 25, Uppsala.

Steve Rogers has worked for Carnegie Museum for 37 years, the first half as preparator in the Section of Birds and the Section of Amphibians and Reptiles, and then for the last 18 years as Collection Manager in both sections. He has only been tangentially involved in exhibits through the years, occasionally completing taxidermy mounts of birds, but has a great interest in the history of taxidermy, exhibits, and scientific preparation. He has collected taxidermy and natural history museum literature for over three decades and has assembled a world class collection.

Rebecca Shreckengast joined the team at Carnegie Museum of Natural History as Director of Exhibitions in 2014. Rebecca worked previously as a designer/detailer and cost-estimator for an exhibit fabrication shop, as a museum exhibit designer and project lead for consultancies in Boston and in Santa Fe, and as a designer and manager for projects executed in state museums and Bosque Redondo Memorial Museum in New Mexico. Most notably, Rebecca served as lead exhibition designer for the North Carolina Museum of Natural History's Nature Research Center project that opened to wide acclaim in April, 2012. She specializes in the collaborative process of exhibition development and in the creation of family friendly exhibit experiences.

She earned her Bachelor of Arts Degree in Industrial Design from Carnegie Mellon University and graduated in 2011 from Harvard University with a Masters of Liberal Arts in Museum Studies.

Dr. Eric Dorfman is Director of Carnegie Museum of Natural History in Pittsburgh, Pennsylvania. Established 120 years ago by Andrew Carnegie, the museum is one of the four Carnegie Museums of Pittsburgh. It is among the top natural history museums in the country, maintaining, preserving, and interpreting a collection of 22 million objects and scientific specimens used to broaden understanding of evolution, conservation and biodiversity. Prior to this, Dorfman was Director of Whanganui Regional Museum in New Zealand.

His PhD, from The University of Sydney, concerned scale-dependent resource use of waterbirds in central and eastern Australia. Before this, he worked on the behavioral ecology porpoise in Monterey Bay, California, through San Jose State University.

Dorfman is an active advocate for natural and cultural heritage, and is an author of popular books on natural history, including Inside New Zealand's National Parks (Penguin 2010). He has also published scholarly papers on museum operations, public programming and the ecology of wetland birds. His most recent work is as editor of Intangible Natural Heritage (Routledge 2012). His most recent book, The Future of Natural History Museums, was published in 2017.

Since 2013, he has been President of the International Council of Museums Committee for Museums and Collections of Natural History (ICOM NATHIST), is a member of the ICOM Ethics Committee and in 2013 published the ICOM Code of Ethics for Natural History Museums. He is also a registered ICOM mediator, chairs the ICOM NATHIST Wildlife Trafficking Working Group and is a member of ICOM's Museum Definition Working Group.

In New Zealand, he was Chair of Visit Whanganui, the regional tourism organization, Vice President of the Whanganui Chamber of Commerce and President of the International Council of Museums New Zealand.

Chapter 3
Developing a Modern Diorama: Changing Habitats in Scotland Since the End of the Last Ice Age

Andrew C. Kitchener

3.1 Introduction

As part of the overall development of a new Museum of Scotland in Edinburgh (McKean 2000), which aimed to tell the history of Scotland from the beginning of its known geological development more than 3 billion years ago to the twentieth century, a small gallery, *Beginnings*, was planned to tell the geological and biological history of what eventually became Scotland.

I was given responsibility for developing the part of the exhibition that dealt with the biological history of Scotland since the end of the last Ice Age, c.11, 750 years ago. A key problem was to try and show how habitats and their associated flora and fauna had changed during the Holocene as the climate warmed. In order to solve this, I decided to develop a large diorama as the principal exhibit within the gallery (Fig. 3.1).

Dioramas, representing natural habitats and their associated fauna and flora, were much in vogue during the late nineteenth century and the first half of the twentieth century. As a child I remember being inspired by the three dramatic dioramas that formed the Rowland Ward Pavilion at the Natural History Museum (these were sadly dismantled in 2004; Morris 2010) and later by the much older and no less remarkable dioramas at the Powell Cotton Museum, Birchington, Kent. I find dioramas compelling museum exhibits because they are dramatic, they place animals and plants in appropriate habitats, which help visitors understand their adaptations, and they are memorable long after the visit, acting as a window on another world, a three-dimensional life-size picture. Therefore, it came as a surprise when some museum colleagues questioned why I should want

A. C. Kitchener (✉)
Department of Natural Sciences, National Museums Scotland,
Edinburgh, UK
e-mail: a.kitchener@nms.ac.uk

A. Scheersoi, S. D. Tunnicliffe (eds.), *Natural History Dioramas – Traditional Exhibits for Current Educational Themes*,
https://doi.org/10.1007/978-3-030-00175-9_3

a

b

c

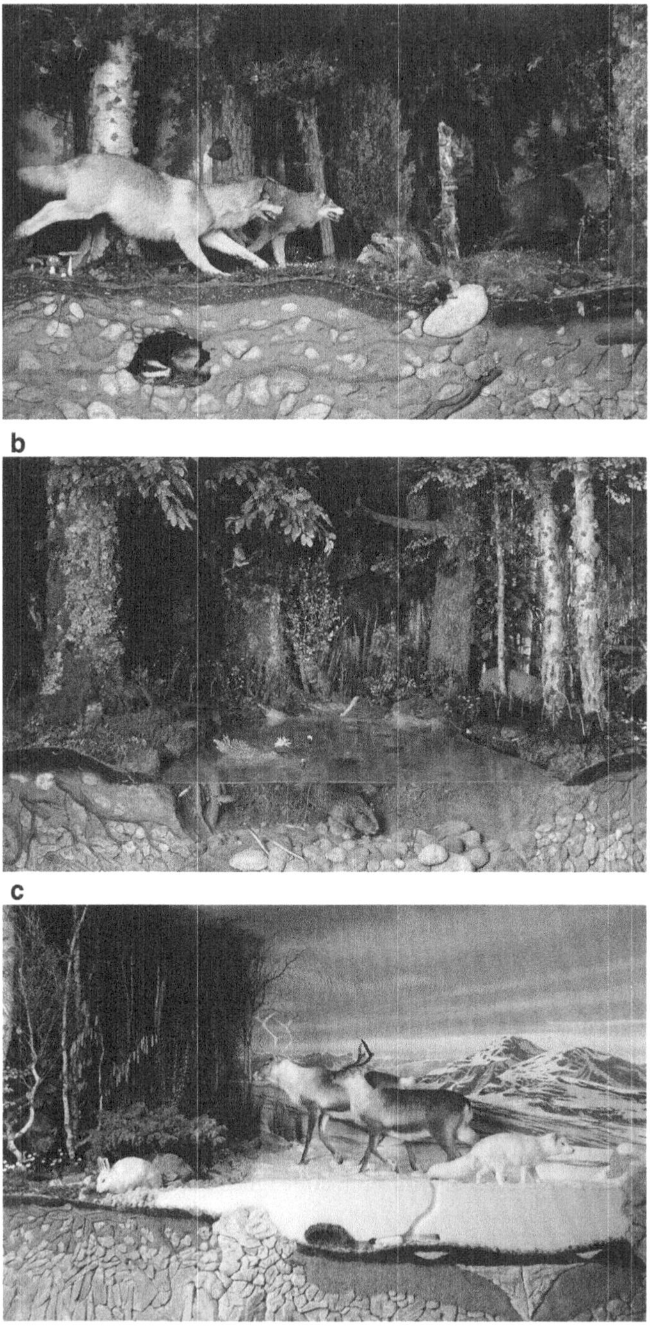

Fig. 3.1 The *History of the Forests* diorama in the *Beginnings* gallery, Museum of Scotland (now National Museum of Scotland) in 1998, (**a**). Caledonian pine forest, (**b**). lowland oak woodland and (**c**). tundra. © National Museums Scotland (Neil McLean)

to use such an "old fashioned" way of interpreting the natural world at the end of the twentieth century and, even more remarkably, why I was allowed to do it! In this chapter I want to show how I developed the ideas that eventually formed the *History of the Forests* diorama in the *Beginnings* gallery of the Museum of Scotland (now National Museum of Scotland) in Edinburgh and the innovations which I believe offer new opportunities of reviving dioramas as a powerful way of interpreting the natural world in the twenty-first century. This diorama has been mentioned in several publications mainly by Tunnicliffe (e.g. Clarke 1998; Reiss and Tunnicliffe 2011; Tunnicliffe 2005, 2006, 2013; Tunnicliffe and Scheersoi 2015), but until now there has been no detailed explanation of its origins and aims. The text for *Beginnings* formed the basis of a linked book by Taylor and Kitchener (2007) that follows the exhibition's themes, including the diorama which is presented in reverse to fit the format of the book.

Although the primary focus of this chapter is charting the development of the *History of the Forests* diorama, I also want to discuss its limited post-opening evaluation and argue for the need for effective and robust evaluation of natural science museum exhibits, including dioramas, if we are to improve their effectiveness in communicating key themes and messages about the natural world.

3.2 Background

A new national museum for Scotland was promulgated and promised as far back as 1953, but lack of funding and political will meant that there were several false starts (McKean 2000). In 1985 the Royal Scottish Museum and the National Museum of Antiquities merged to form the National Museums of Scotland (now National Museums Scotland), which revived the will to provide a home for the national collections on one site in Chambers Street and led subsequently to the birth of the Museum of Scotland Project in 1990 (McKean 2000). As Principal Curator of Mammals and Birds at the time, I was asked to develop an exhibition proposal for the history of Scottish wildlife. My first attempt in March 1990 proposed a traditional museum exhibition, comprising a series of cased exhibits of individual specimens. However, there was a small glimmer of what was to come; I had proposed that visitors....

> would enter a cave and see subfossil bones of Ice Age and post Ice Age animals from the Inchnadamph caves *in situ*, including signs of human habitation. As they leave the cave they would come across an Ice Age diorama. This could either simply be a painting or it could contain specimens of reindeer, lemmings, arctic fox, mountain hare and ptarmigan.
>
> The contrast between the sparse fauna of the tundra and the diverse fauna of the post glacial vegetation would be shown in two other dioramas of a typical Caledonian pine forest and a lowland deciduous forest.

At that time I imagined that these exhibits would be actually quite small and separate from each other, and would only show a limited range of the species that were extant at those times. These small dioramas would have a few small key specimens and may have relied on painted images of some large animal and plant species. As I had initially envisaged them, these would not have had the same impact as a full-scale large diorama as developed in many museums around the world.

3.3 Exhibition Development

The Museum of Scotland was designed by architects Benson and Forsyth and was intended to show the history of Scotland from its earliest geological origins almost 3000 million years ago until the twentieth century. The history of the changing landscapes of Scotland and its wildlife were combined into a single gallery, *Beginnings*, which was developed jointly with my colleagues in the then Department of Geology, firstly Dr. Ian Rolfe, Keeper of Geology, and latterly Dr. Michael Taylor, Principal Curator of Vertebrate Palaeontology, and Sue Mitchell of the then Education Department. The key transition between our respective sections of the gallery was the Last Glacial Maximum – the peak of the last Ice Age about 21,000 years ago. Jointly we developed a common approach to presenting and interpreting the themes within the exhibition:

Reconstructions would be used to aid in visual understanding of the key messages we wished to convey. These reconstructions, whether paintings, models, vignettes or dioramas, would be presented alongside actual evidence, including rocks, minerals, fossils, pollen grains, etc. In that way visitors would be able to make a direct link between the actual physical evidence of past life or geological events and our interpretations of them in terms of what animals, plants, habitats and landscapes probably looked like.

Our major limiting factor was space. Despite the aim of telling a story approximately 3000 million years long, we had ultimately only 300 m^2, i.e. approximately 10 million years per square metre. Key messages were developed for the entire exhibition in a hierarchical way. This not only provided a clear structure against which we could develop the exhibition elements, but also ensured that the storyline was sufficiently honed and streamlined to fit the limited available space. Exhibition design was by TPS Dangerfield (later TBV Dangerfield) during 1995. It soon became clear that constraints of space meant that instead of three separate dioramas for the *History of the Wildlife* section, we would have to combine the three habitats (tundra, Caledonian pine forest and lowland deciduous forest) into a single diorama. This proved advantageous because we now had one large semicircular diorama, almost 15 m long by 5.5 m deep, to fill, and because it also provided an opportunity to develop a chronology and show a direct transition between these habitats that would not have been possible with three smaller, separate, dioramas.

3.4 Principles Behind the Design of the History of the Forests Diorama

As the biggest and most important exhibit in the *History of the Wildlife* section of *Beginnings*, it was important that the diorama should be successful in conveying its key messages, which are:

Key Message

There were dramatic natural changes in the fauna and flora after the last Ice Age and before the arrival of Neolithic farmers.

Subsidiary Messages

The first colonists – animals and plants of the tundra – were replaced by trees and forest animals as the climate warmed.

The tree cover changed from willow/birch-dominated woodland to Scots pine-dominated forest in the Highlands and oak/hazel-dominated forest in the lowlands.

The fauna changed as the flora changed, as indicated by subfossils. Natural extinction of Arctic species coincided with natural colonization by forest species from the south.

Some animals and plants were generalists that were able to survive in a range of forest habitats but others were dependent on specific conditions.

Some animals have changed in size over time, reflecting environmental degradation caused mainly by human factors.

I developed the following key principles for the exhibition as a whole, but the diorama in particular:

- *Realism*. As far as possible the diorama should have realistic representations of the habitats it displays. This required a completely new exhibit, including taxidermy, models, preserved plants, scenic backdrop, soil profiles, etc. For example, the collared lemming, *Dicrostonyx torquatus*, nest under the snow is a real nest from Greenland, albeit from a closely related species, *Dicrostonyx groenlandicus*.
- *Interaction*. Many dioramas show animals just standing or looking back at the observer, or involved in incidental behaviours, such as walking, looking, scratching or (commonly for carnivorous mammals) snarling. I wanted to show animals displaying dramatic natural behaviours and interacting with each other, which highlighted key adaptations, behaviours or interactions between species, which visitors could easily interpret by looking at the diorama. For example, two wolves, *Canis lupus*, are chasing a wild pig, *Sus scrofa*, which has startled a swimming beaver, *Castor fiber*, which has slapped its tail on the water to alert its fellow beavers of possible danger (Fig. 3.1a, b). Or an Arctic fox, *Vulpes lagopus*,

which is taking an interest in visitors as it trots along in pursuit of lemmings (under the snow), has surprised a pair of reindeer, *Rangifer tarandus*, feeding on reindeer moss (Fig. 3.1c). Or a lynx, *Lynx lynx,* is stalking a mountain hare, *Lepus timidus*, but a wood mouse, *Apodemus sylvaticus*, has been disturbed and is running, mistakenly, for its life (Fig. 3.1c).

- *Time*. I wanted to show not only the big time changes from tundra at the end of the last Ice Age to the peak climax forests of up until about 5000 years ago (when Neolithic farmers began to change the natural landscape dramatically and permanently), but also on a seasonal and daily scale. Therefore, the tundra represents winter, the transition spring, lowland oak woodland is summer and Caledonian pine woodland is autumn. I also wanted a dawn-to-dusk light cycle, in order to allow higher peak light levels than if the diorama were lit permanently, and also to try and minimise reflections from the cases opposite, which were unavoidable, owing to lack of space. However, this light cycle was not achieved until 2012 when the diorama was extensively renovated after a pest outbreak.
- *Size*. Big animals tend to be the focus of dioramas and hence dominate them. This disadvantages smaller specimens, which tend to be overlooked. I decided that the two biggest animals, an elk, *Alces alces*, and a brown bear, *Ursus arctos*, would be hidden at the back of the diorama, so that they would not dominate it (Fig. 3.2), but also provide elements of surprise as they appear in the gloom at the back of the scene. This allowed us to pack the front of the diorama with very small specimens (flowering plants, fungi, slugs, snails, small birds and mammals) and the middle with small to medium-sized animals. By situating the

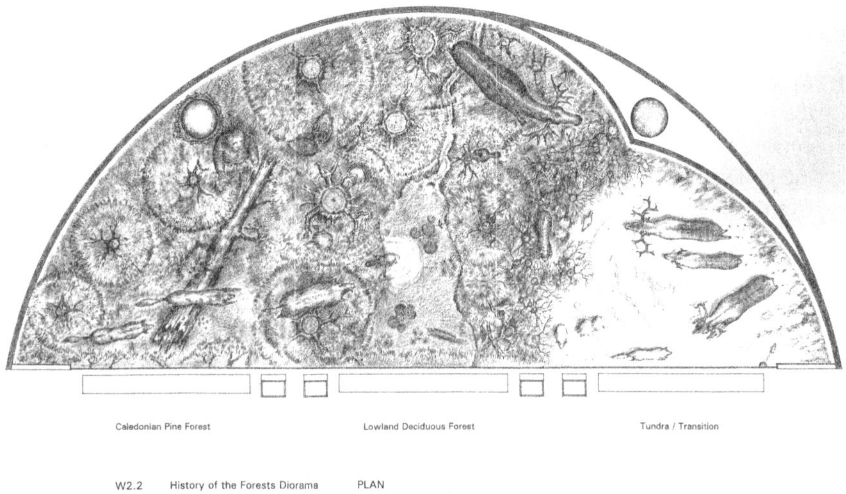

Fig. 3.2 Original plan view of the diorama from the *Beginnings* Phase 1 Report by TBV Dangerfield, November 1995, showing large mammals at the back and smaller animals in the foreground. N.B. One reindeer was not used owing to lack of space. © National Museums Scotland

undulating ground level of the diorama 60–70 cm above the floor level, which allowed large specimens to be fixed into place from below the staging, it also allowed us to reconstruct appropriate soil profiles and features, e.g. an ice wedge (albeit somewhat truncated), and burrows showing animals under the ground, including earthworms (Fig. 3.3), badgers, *Meles meles*, a mole, *Talpa europaea*, and a collared lemming.

- *Ecology*. Although primarily intended as a chronological diorama, I also saw the potential for its use in interpreting ecology. Interactions between predators and prey, and herbivores and plants, demonstrate simple food webs, trophic levels, ecological niches, etc. For example, a sparrowhawk, *Accipiter nisus*, is pursuing a robin, *Erithacus rubecula*, in the oak woodland (Fig. 3.4), just as can be seen in gardens in Edinburgh today. Crossbills, *Loxia scotica*, are using their curious bills to prise open Scots pine cones to feed on the seeds within.

In cases opposite the diorama presents the evidence for how we know about the reconstruction we have made. There are fossils of specimens from Scottish localities, which are represented in the diorama, as well as exhibits which describe methods for reconstructing the past, such as radiocarbon dating and pollen analysis. This includes a complex interactive which allows visitors to input real data for radiocarbon levels measured in specific fossils to estimate radiocarbon and calendar dates for particular Holocene mammal and plant specimens.

The diorama required extensive background research, most of which I carried out during the development phase. On the botanical side I was assisted by Professor Jim Dickson of the then Department of Botany, University of Glasgow. Taxidermy was done in-house by Phil Howard and Peter Summers, but the habitat reconstructions were by external contractors, Derek Frampton and Steve Massam, and the painted backdrop was by Ted Michell. Mr. Johnnie Grant of Rothiemurchus kindly gave permission for the collection of plants, mosses, soil and other natural materials, which were preserved and freeze-fumigated prior to installation.

Fig. 3.3 Earthworm in its burrow in the section through the soil of the diorama. © National Museums Scotland (Neil McLean)

Fig. 3.4 A sparrowhawk in pursuit of a robin. An ecological interaction. © National Museums Scotland (Neil McLean)

The diorama was constructed off site by Frampton and Massam, and the elements were brought together for assembly in the gallery shortly prior to the exhibition opening in November 1998.

A Finishing Touch One of the most popular and unplanned additions to the diorama was made by Derek Frampton and Steve Massam on its completion. They constructed a small snowman, which they placed in the doorway to the tundra end of the diorama, so that it can only be seen by visitors looking to the right at the extreme end (Fig. 3.5). Although intended as a humorous postscript and despite its cryptic positioning, it attracts much attention (and raised some concern amongst our visitors when it accidentally toppled over a few years ago) and could be used as an interesting discussion point about the role of people, who are missing from the diorama, in the Mesolithic landscape and ecology of Scotland. The exclusion of people was deliberate, partly because the Mesolithic human story is told in the adjacent *Early People* gallery (Clarke 1998), and partly because we would have no way of convincingly knowing what clothes people would have worn and we would have risked giving a false interpretation. Given that human impacts were limited at this time, and we wanted to show pristine habitats, it was decided to exclude people.

Interpretation Interpreting dioramas can be problematic. In the first instance visitors engage visually with the dramatic scene, looking at the "big picture" and the specimens therein and possibly identifying them based on previous knowledge and experience. However, visitors often want to engage with some form of formal

Fig. 3.5 The snowman. © National Museums Scotland (Andrew Kitchener)

written interpretation to confirm their initial identifications or answer questions they may have. For most dioramas, either there is little or no interpretation, perhaps a general panel at each end with silhouettes of the main protagonists labelled for identification, or at the other extreme there is too much interpretation, with a forest of labels that obscures the diorama, and which visitors give up reading because they are overwhelmed by the labels or the labels are often not placed sufficiently close to the specimens they are identifying.

I wanted to try a different approach, so instead we developed three touch-screen interactives, which each display a photograph of the particular habitat that corresponds to that part of the diorama in front of them (Fig. 3.6a). Simply touching the animal or plant in the photograph brings up a pop-up label identifying it (Fig. 3.6b, c). Another tap brings up more detailed information such as a small amount of general text about the species in Scotland, a distribution map, length, diet, habitat and range (Fig. 3.6d). In my view, the immediacy and directness of this approach encourages visitors to interact with the diorama, rather than discouraging them by forcing them to make two or three different actions to identify each specimen. Visitors do not need any prior knowledge to quickly and accurately identify the specimens they can see and discover further information if they wish. These interactive screens are physically compact and hence do not disrupt the visitors' view of the diorama, but they are packed with information if required. The touch-screen programmes were developed by IBIS at the University of Nottingham; an important feature of the content was the explicit archiving of the programme to allow us to edit and update it in-house as required.

3.5 Problems

Two main problems can occur when taxidermy is displayed; light levels cause fading of specimens, and ingress of pests and dust cause direct damage and even total loss. In addition, fluctuating humidity levels may cause specimens to split and crack. In the case of the *History of the Forests* diorama, light levels and humidity were controlled within acceptable limits, but I was always concerned about ingress of insect pests. Doors and glazing appeared to be well sealed, and filters on doors allowed some degree of air circulation without risking pest ingress. The main

Fig. 3.6 How the touch screen interactive (**a**). shows the scene in front of the viewer, can be touched (**b**). to identify the animal or plant (**c**). and provides further information appears (**d**), if required. © National Museums Scotland (Andrew Kitchener)

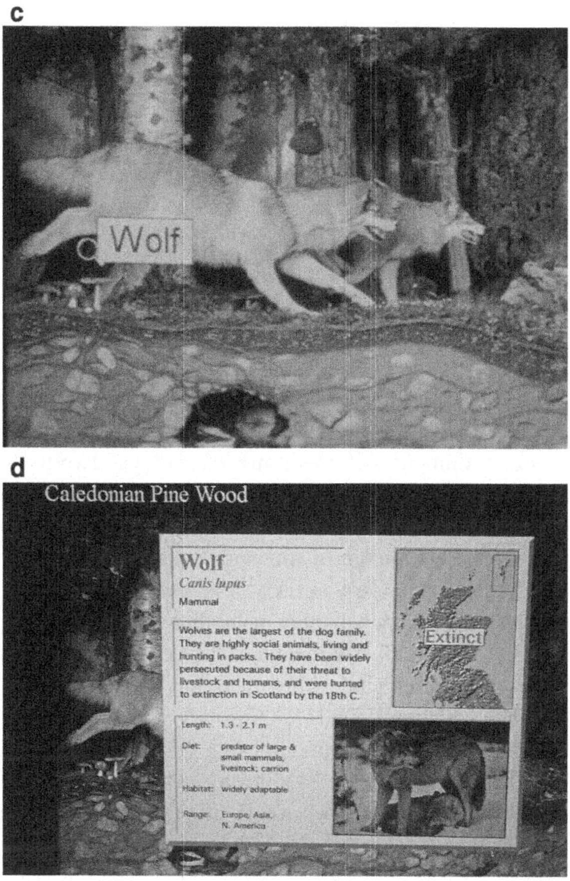

Fig. 3.6 (continued)

lighting area at the top of the diorama was supposed to be contained in a glazed light box, which allows external access for changing light bulbs and other maintenance and repairs. However, the structure of the down-lights at the bottom of the lighting box to light up the soil profile and small specimens at the front of the diorama was such that there was no effective seal between the light box and the diorama exhibits. Moreover, over time visitors picked away the silicone seals between the three large pieces of plate glass that comprise the window into the diorama. In late 2012 it was noticed that there was an infestation of clothes moth, *Tineola bisselliella*, in the diorama. The decision was made to try and reseal the diorama and assess the damage to specimens, replacing any that were beyond repair or conservation. If damage were extensive, there was a real risk that we would have had to dismantle the diorama and install a different exhibit that attempted to convey similar messages. Several alternatives were discussed, but costs were prohibitive and the logistics were difficult, so it was decided to try to clean, reseal, relight the diorama, and repair or replace any specimens, treating all taxidermy with insecticides. The moth problem was controlled using a fogging treatment six times during the following year, but these did not eliminate the moths.

Treatment and renovation of the diorama were carried out during February and March 2014. Damage assessment showed that the elk was the main focus for moth activity and, although it was no longer as pristine as it once was and it is in a delicate state with hair liable to fall out when touched, the decision was made to try and treat it and other taxidermy in the hope of eliminating the moth infestation and preventing further pest outbreaks. Seals on doors and between glazings were renewed; silicone seals were replaced by aluminium seals to prevent visitor attack. The lighting system was replaced with low-energy LED lighting and fibre-optic lighting to minimise the build-up of heat in the diorama, so that the filters in the doors were closed up, thereby eliminating a further possible route of pest infestation. The lighting is also now completely sealed into the lighting box, so that the diorama is sealed off from the ambient environment. All taxidermy was sprayed two to three times with Johnson's Cage 'n' Hutch aerosol insect spray, which is a permthrin-based insecticide, which also contains an insect growth regulator that prevents eggs hatching and larval development. The advantage of this spray is that it does not disrupt fur or feathers when applied and dries very quickly. It is also highly effective – when a red deer, *Cervus elaphus*, stag on open display in another gallery was infested by clothes moths, a quick spray of Cage n' Hutch caused larvae to leave the specimen immediately and die very soon after, since when there has been no further moth activity. Powdered insecticide (Killgerid and K-Orthrine) was also spread through the ground work and a final fogging treatment, using ULV500, was carried out prior to the gallery reopening in March 2014. Monitoring of moth traps after the restoration of the diorama showed three moths two weeks after and no activity since. So at the time of writing, our attempt to eliminate clothes moths from the diorama has been completely successful. However, monitoring continues. There was also evidence for house mouse, *Mus domesticus*, activity, but this was not recent and probably dated from the early stages of the later Royal Museum Project, when owing to the extensive building works, some house mice were able to enter the museum complex.

3.6 Post–opening Evaluation

Upon the opening of the gallery, the immediate reaction of visitors to the diorama was very positive. Anecdotally, levels of interaction with touch screens were high and small children could be seen pulling their parents down to the floor to see the slugs and snails at knee height, or to peer into the badger's sett. However, for several years there was no formal evaluation of the effectiveness of the diorama.

In April 2003 Sue Dale Tunnicliffe carried out an evaluation of the *History of the Forests* diorama by looking at what visitors understand from it and how they interact with it. This evaluation was limited to a single day, so the sample size was limited (n = 25), but the gathered data were first evidence of how visitors interact with this

Table 3.1 Categorisation of responses to *History of the Forests* diorama, *Beginnings* gallery, National Museum of Scotland (Tunnicliffe 2005)

LEVEL OF RESPONSE	Subcategory of response	Example of activity
First LOCATE	1.1 Spontaneously	Pick out things in dioramas e.g. wolf, fungus, snow
	1.2 Assisted – using signage	See things on interactive and then locate them in diorama
Second IDENTIFY BY NAME	2.1 Spontaneously	From own knowledge or that of companion
	2.2 Assisted	From signage or guide
Third DESCRIBE	3.1.1 Physical aspects	From own experience and knowledge- e.g. wolves chasing wild pig, beaver swimming, 'fox' is white
	3.1.2 Behaviours	
	3.2 Scene	It's winter because there is snow'
Fourth INTERPRET	4.1.1 The visitor's story using own knowledge	E.g. 'It's telling the story of the seasons"
	4.1.2 Visitors story using museum information/ message	'It's about the changes in the landscape'
	4.2 The museum's story	The change in flora and fauna since the ice age

diorama. Depending on which end of the diorama they started and with their permission, visitors were either accompanied and their conversations were recorded (in writing or electronically), as they walked the length of the diorama, or they were asked about what they had just seen. Some demographic data were recorded including age, sex, and from where they had come to visit the museum.

Some visitors entered the exhibit 'the wrong way', i.e. from the chronologically more recent end (Caledonian pine forest). Few read the information panels, which provide brief information about the context for the diorama. More interacted with the touch screens in front of each of the three sections of the diorama.

Subsequent analysis of the recorded data revealed a number of levels of interpretation and usage of the dioramas, which are summarised in Table 3.1 (Tunnicliffe 2005). The youngest children tended to interact with the diorama at the lowest level and levels of interaction tended to increase with age. The simplest level of interaction was locating specimens within the diorama and the next level was to try and identify them using their own or a companion's knowledge or by using the touch screens. The third level of interpretation was describing what was happening in the diorama, i.e. the interactions between animals, plants and the environment in which they are immersed, which was often based on visitors' own experiences. The final level of interpretation was an understanding of the overall message or story behind the diorama, such as changing seasons, or the intended story of the history of climate and environmental change following the end of the last Ice Age. Overall, we are encouraged that visitors interacted with the diorama directly or via the touch screens to interpret the diorama in the ways that we envisaged. It is evident from the content of the conversations that the diorama is a valuable resource for science education and many activities could be developed, which would enhance the learning experiences of children and adults. However, it is also clear that more extensive evaluation would contribute invaluable data about the impact of this and other dioramas in conveying important messages about the natural world.

Acknowledgements I am grateful to Sue Dale Tunnicliffe and Annette Scheersoi for inviting me to write this paper and to my colleagues Mike Taylor and Xerxes Mazda for their comments on an earlier draft, but above all I am most grateful to the many colleagues and contractors, mentioned above, who worked with me to make this extraordinary diorama even better than I thought it could be and helped save it when pest attacks seemed to have led to its downfall.

References

Clarke, D. V. (1998). New things set in many landscapes: Aspects of the Museum of Scotland. *Proceedings of the Antiquaries of Scotland, 128*, 1–12.

McKean, C. (2000). *The making of the Museum of Scotland*. Edinburgh: NMS Publishing.

Morris, P. (2010). *A history of taxidermy. Art, science and bad taste*. Ascot: MPM Publishing.

Reiss, M. J., & Tunnicliffe, S. D. (2011). Dioramas as depictions of reality and opportunities for learning in biology. *Curator: The Museum Journal, 54*(4), 447–459.

Taylor, M. A., & Kitchener, A. C. (2007). *Scotland's beginnings. Scotland through time*. Edinburgh: National Museums of Scotland.

Tunnicliffe, S. D. (2005). What do dioramas tell visitors? A study of the history of the forests diorama at the Museum of Scotland. *Current Trends in Audience Research and Evaluation, 18*, 23–31.

Tunnicliffe, S. D. (2006). The importance of research to biological education. *Journal of Biological Education, 40*(3), 99–100.

Tunnicliffe, S. D. (2013). Animals and plants in natural history dioramas in museums: Specimens or objects? *Journal of Biological Education, 47*(4), 189–191.

Tunnicliffe, S. D., & Scheersoi, A. (2015). Dioramas as important tools in biological education. In S. D. Tunnicliffe & A. Scheersoi (Eds.), *Natural history dioramas. History, construction and educational role* (pp. 133–143). Dordrecht: Springer.

Dr. Andrew Kitchener is Principal Curator of Vertebrates at National Museums Scotland. He has worked on many permanent and temporary exhibitions since starting work at the museum in 1988, including *World in Our Hands* (1992), *Bird Biology* (1994), *Beginnings* (1998), *Cats … the ultimate predators* (2004), *Animal World, Animal Senses* and *Survival* (2011). His latest exhibition is *Monkey Business*, which opened in December 2016 and is now touring museums around the world. Andrew has many research interests, including geographical variation in vertebrates, hybridisation, effects of captivity on endangered species, zooarchaeology, functional morphology and biomechanics, and he is an Honorary Lecturer at the University of Glasgow and an Honorary Research Fellow at the University of Edinburgh.

Chapter 4
Building and Maintaining Dioramas – A Field Report

Eirik Granqvist

4.1 Building Dioramas

At the international Directors' Forum on the Future of Natural History Museums, that was held at the Shanghai Science & Technology Museum in 2009, one of the main topics discussed was dioramas. It was stated, that there was no better way to teach the general public about the natural world than with using dioramas: A diorama can show a precise moment in nature with animals, plants and geology in a typical landscape with the endemic species of any place in the world. Therefore, old existing dioramas should be maintained and when needed restored and new dioramas should be built. It must not be forgotten that a diorama with its composition, its pieces of the art of taxidermy and its background painting is itself a piece of complex art and should be understood as such.

Thus, a natural history diorama is a very valuable tool for education purposes. When standing in front of a diorama, the first thing one notices is the impression of a beautiful picture. You are very close to the animals and the plants and what is happening in the diorama. While observing the scene that is shown, you learn without thinking about it.

Creating a diorama is an incredibly complex, artistic process, which demands a lot from its creator. Such a person needs to be both, a very good artist and sculptor but also a competent scientist who knows the appropriate geography. The creator has to be aware not only of the various specialist techniques required in the construction but also the maintenance after its completion. It is increasingly difficult to find such expertise and in turn train the next generation.

The process of constructing a diorama is complex: Initially, once a potential location has been identified, the space that is available for the diorama must be measured. Then the appropriate biotope must be identified and it must be decided

E. Granqvist (✉)
Borgå, Finland
e-mail: jinsong.guan@gmail.com

A. Scheersoi, S. D. Tunnicliffe (eds.), *Natural History Dioramas – Traditional Exhibits for Current Educational Themes*,
https://doi.org/10.1007/978-3-030-00175-9_4

which time of the year it will reflect. In accordance with such information, the specimens that are required must be sourced. The animals must have the right consistency and colour of fur. The plants needed must be chosen, and specimens must be available at the appropriate developmental stage and colour. It is essential, that an accurate representation is constructed and a true depiction is displayed!

To give some concrete examples, the following section will explain the construction of some dioramas that existed in the Zoological Museum of the University of Helsinki.

4.1.1 The Rut-Fight European Elk Diorama

Selecting the Animals The first example is the diorama showing a rut-fight between European elks in September. It is important to create authentic dioramas in order to present the real-world situation as accurately as possible. I had indeed found the place where such fights had occurred and I had heard fighting elks, but they were too shy to let me come close enough to watch their fight. Moreover, during the rutting season, I tried to call the bull by making a noise simulating the call of an elk-bull myself. The bull responded but came alone, believing that I was just another bull coming to steal his female.

In the American Museum of Natural History in New York, there is a famous moose fight that could have served as a model. However, firstly I did not want to copy this and secondly the behaviour of an American moose and a European elk may not be identical. When I found some good photographs of a rut-fight between European elk-bulls in Sweden, I could finally start planning the diorama. I wanted very large bulls and an old and big female for my diorama. The European elk population in Finland is now large and several ten thousand elks are culled by hunting every year. The elk has become an important source of meat for humans. Seen from that point of view, there was not a problem to acquire suitable animals. However, I wanted big and beautiful specimens. Furthermore, it was crucial that the bodies had been handled correctly when bringing them out from the forest, including being skinned and measured. Thus, I hunted the elk myself and took care of the specimens. It took 3 years before I had found the right animals (Fig. 4.1). I will not go into detail about the process of taxidermy, but just give a few examples.

Taxidermy The elk's body was built on a sculpture frame, made of the original leg bones, wood and chicken wire with burlap and plaster (Fig. 4.2). On this frame, modelling clay was placed to sculpt the body with its muscles. The form was then cast in plaster and burlap. The cast was hollow like an eggshell. The skin was then glued onto the artificial body.

When looking at the diorama (Fig. 4.1) we recognise two jays to the right. They are not just decoration, but they inherit a special function in the diorama: Jays often make noises to attract large carnivores to the place. If they succeed, they might get

Fig. 4.1 The European elk-diorama as it was shown in the Zoological Museum of the University of Helsinki

Fig. 4.2 The elk sculpturing frame waiting for the clay

their share of meat if an elk is killed and eaten. When showing birds in flight in a diorama, it is very important not to hang the birds on a nylon thread, because a thread can always be seen. Therefore, the bird should be attached to an iron thread from the wing, the tail, the leg or the body to a branch, a stone or to anything else in the way, so that we cannot see the thread but have the feeling that it is just flying freely.

Using Plants Plants used in dioramas can be preserved in some way, or modelled with other materials. The moss shown in the diorama had to be taken in complete pieces in the swamp and then brought out in layers, isolated from each other with plastic sheets in plastic bags. Back at the museum the moss parts had to be put on plywood pieces and brought carefully arranged into the freezing room for a couple of weeks, for them to freeze dry. Then the moss had to be placed in the diorama and sprayed with shining varnish, in order to give it the impression of moisture. In addition, modelled and well red-coloured cranberries as well as fallen, yellow autumn leaves were placed in the moss. The leaves had been pressed flat, like for a botanic collection, then spray-coloured with latex paint. The leaves of aspen and birch on the branches were treated in the same way. Small and not very deep holes were bored to fix the branches and the leaves.

To the right side of the diorama, there is a small pine tree with the bark scratched away, where the elk-bull has scratched away the so-called velvet skin, by cleaning its antlers, when it was fully grown. The bark free area had to be coloured back in order to appear fresh. Hence a layer of shellac had to be sprayed on to give the impression of freshness.

The branches of the pines and spruce had to get their special treatment, too. They had to be put in a mixture of half water, a quarter of alcohol and a quarter of glycerine for at least 2 weeks. Then they were taken out and dried. When treated in this way, the needles will not fall off but will turn dark brown. The branches should be hung up for drying and the needles spray-coloured again with a shiny oil paint.

In another very small diorama a northern birch mouse is climbing on the branch of a birch tree. In this diorama, commercial plastic birch leaves were used that were reshaped to the shape of a real northern birch. The same method was used for the leaves of an apple tree in a diorama with a garden dormouse. As I could not find any commercial apple tree leaves, I took the artificial rose leaves and reshaped them. The plastic stalks of the leaves were melted over a flame of an alcohol laboratory lamp and glued to the branch with the melted plastic itself.

The Viewer's Perspective Care is needed in preparing the diorama for viewing. Notice in Fig. 4.3 that the window is leaning towards the visitor to avoid reflection. If the diorama glass is placed upright, visitors will see the reflections of themselves almost like in a mirror. However, if the glass is leaning towards you, the reflections are directed downwards.

The window is placed 70 cm above the floor on which the visitor is standing and the ground in the diorama is about 50 cm higher than the floor. The ground of the diorama is higher because in real life you see the animals at distance. About the same angle for viewing should be recreated, even though the animal is viewed at a very close distance in the diorama. Hence, the ground of the diorama will be arranged higher than the floor on which the visitor is standing (see Fig. 4.4). In building the foreground a bit higher than the rear of the artificial landscape, you get

Fig. 4.3 The author in front of his diorama

Fig. 4.4 Recreating a feeling of distance

a more accurate feeling of distance for avoiding the box-impression. If the animal and the artificial ground are at the same level as you are standing, it will give you the impression of looking at a box. The visitor will lose the illusion.

Fig. 4.5 The artist John Grönvall is painting the background for a Lapland diorama

The Design of the Diorama Background The background of the diorama should be curved because you cannot paint away corners. The best is a curved background that will help you to get the feeling of space, without actually using such a large area. You should also not place the animals along the walls of the diorama because that would only increase the impression of a box. The animals should be placed in the middle and fairly close to the diorama window.

Painting diorama backgrounds is a difficult skill. In the Lapland diorama (see Fig. 4.5), the artificial landscape shows the slope of a mountain. On the round and cupola shaped background, the landscape must be painted quite low, in order to create an authentic image of the actual place. The height of the horizon plays a key role: to create the impression of more depth, the horizon has to be placed a bit lower.

Size of the Diorama Window Quite often we can see dioramas with very small windows. These are made with the idea in mind that the visitor should have to look through a small hole, for discovering what is really shown in the diorama. I feel that this is just frustrating. The visitor should be able to stand at a certain distance and admire the diorama with its content and just dream. Therefore, the window of the diorama should be large enough to see the whole animal or animals, without any problems. If you have a large diorama, you can leave a space between the background painting and the artificial ground. That gives a better feeling of depth and the possibility for the museum staff to work inside the diorama for correcting things, or just for giving a normal service.

Lighting Dioramas are generally illuminated using neon light tubes. So called warm light neon tubes create a perfect light for summer or tropical landscapes, and their bleaching effect on the items is not important. The so called white light tubes

Fig. 4.6 Diorama with Arctic scavengers (a golden eagle at a reindeer cadaver together with ravens, a wolverine and an arctic jay)

give a cold, bluish light. These tubes are excellent for winter landscapes, since they give a feeling of coldness. They are known for bleaching more but since the winter animals are often white, or light in colour, or just black like ravens this is not a big problem. The neon tubes in the diorama should be hanging from the ceiling. The light should be strong in the foreground of the diorama and less towards the background. If the diorama is larger, then you might have to place additional neon tubes on the floor between the artificial landscape and the background painting, thereby avoiding shadows from the animals on the background. If the feeling of night time is required, the neon tube may be painted with a small amount of latex colour, just leaving the necessary light to come through.

4.1.2 The Arctic Scavengers Diorama

Animal Placement When you create such a composition (Fig. 4.6), the animals and birds are not merely placed randomly but with a carefully elaborated plan. The wolverine is carefully sniffing the cadaver and waiting how the eagle is going to react. The eagle is doing the same towards the wolverine. One raven is about to snap the eagle's tail, giving the raven the possibility to get in front of the eagle and grab a piece of the cadaver. Three other ravens are feeding and the fourth one, behind the reindeer's leg, is looking to get his share, while another raven is just about to land. The arctic jay has carefully snatched a small piece for himself, while the ravens are too busy to look at him.

Using Snow The snow in a diorama was intended to appear as fine, cold winter snow. Hence, in this diorama snow was imitated by using fine white sugar on a white latex painted layer of chicken net, with burlap and plaster. The sugar glitters

Fig. 4.7 Fox den diorama

a little in the artificial light and gives a realistic impression of real snow. On this sugar snow and on the reindeer cadaver, some artificial 'Christmas tree snow' from a spray bottle was applied.

The effect of spring winter snow, which had already partly melted but then had frozen again, can be created by using sugar on which you carefully spray a little water and let it dry. When well dried, just spray a little bit of 'Christmas snow' from a spray bottle on this snow and on the plants.

4.1.3 Very Small Dioramas

In some cases, when the subject is focused on a very small object, dioramas are quite small. Very small dioramas should be placed higher up but still low enough for children to have a good viewing experience. I used to place them in a way that the bottom of the frame is approximately 1.10 m above the floor.

Some small dioramas can be displayed to utilize small spaces. A fox den diorama (Fig. 4.7) was placed in a very narrow space between the wall to the left and a concrete pillar to the right. The wall and the pillar were hidden in granite rocks made with chicken net, burlap and plaster with a thin layer of a mixture of wood-pulp, plaster and carpenters glue. When the thin layer was put on, the structure was just pressed in by using a natural, wet sponge. When dry, the stone was painted with a paint mixed with thin shellac, to which was added colour pigment. Then the surface was swept over with a wet cloth, for taking away the colour of the upper parts and leaving it darker in the lower parts of the surface. Between, there was just the space for having the laying fox-mother watching her youngsters in front of the entrance to the den. The entrance begins just in the front of the youngsters and follows down under the window frame at the front.

4.2 Collecting Materials – The African Wildlife Dioramas

Animals are not only sourced from the wild and taxidermy does not automatically present a threat to wildlife. Specimens can be sourced from road kill, natural deaths and zoo deaths.

In 1973, however, it was decided that a trip should be made to the Kalahari Desert and the Okawango swamps in Botswana, to collect material for an African themed diorama hall in the Zoological Museum of the University of Helsinki.

When arriving in Johannesburg in South Africa, I started collecting by buying the skins of two Blesboks in a butcher shop, where I also skinned and measured them. On the roof of a very cheap hotel, I salted the skins to preserve them. These became the first specimens for the African Hall in Helsinki.

As soon as we received the hunting permits, we started our expedition into the Kalahari Desert. At that time, it did not have any proper topographic maps.

Sourcing for biofacts from the wild, has proven to be useful for future uses. I took one sausage tree fruit with me from Okawango and made a silicone mould out of it, so that I could later cast as many as I wanted.

In the desert, we also collected other plants for the dioramas and insects, as well as birds. All was accurately noted and photographed. Back in the museum, the diorama should be authentic. This was not a conceptual diorama but an accurate representation of a real place. The Kalahari Desert is in fact not a naked desert: It has lots of grass, bushes and trees. Therefore, the fauna is very rich. In Kalahari, there are no oases and the region gets rain only during about 2 months of the year, which gives plants the possibility to grow. In the desert, there are so-called salt pans with a layer of white salt and when it rains these saltpans often have a thin layer of water for some days, which appears as if they were lakes.

In the background painting of the Kalahari diorama, (Fig. 4.8) one can see the Lothlake saltpan with a natural assembly of game of different species in the morning. In reality, this is also a piece of history because you can hardly see such clusters anymore. For protecting the domestic cattle, in the interest of avoiding diseases like foot and mouth disease in this country, where the meat export is a very important and part of the national income, the country has been divided with cattle and game fences that stop the natural migrations of game. This has diminished the game up to 80%. There remains a lot of game to be seen, but no longer the tremendous herds as they existed in 1973.

Such dioramas with scenes of the past form a historical record and are part of the cultural and natural legacy of areas. Even if some protected areas like Moremi and Chiefs Island still keep very strong wildlife populations, the old paradise is gone, and my constructed dioramas became historical evidences. Unfortunately, those Dioramas are now gone as well.

Fig. 4.8 African Kalahari dioramas under construction (note how the glass of the lion diorama is leaning towards the visitor to avoid reflexes)

The Tsessebe antelope in the Okawango diorama (Fig. 4.9, dark animal to the left) was shot and collected by a famous Africa traveller beginning 1900. The trophy was a world record at that time and the taxidermy was carried out by Rowland Ward in London. I decided to use this famous specimen in the diorama.

For the construction of the Kenya lion diorama (Fig. 4.10), the lions as well as the hyena came from a zoological garden. I had been in Kenya and documented the biotope accurately. The grass was authentic grass from there, as well as the grass-hoppers. In the acacia are hanging weaver bird nests from Kenya. The acacia leaves are commercial mimosa leaves that were cut smaller and shaped like acacia leaves. The branch attached to the upper frame of a diorama, gives a better feeling of depth.

4.3 Maintaining Dioramas

In my opinion, the greatest danger to dioramas is museum staff, who are not competent or untrained in diorama construction and maintenance, which is time-consuming and expensive. Another danger are other living things, such as insects. There are some simple things that can be done to avoid insect damages in dioramas. At first, all the taxidermy items should be treated with Eulan, a chemical used mainly in the wool industry. It is not poisonous but it changes the proteins in a way that insects cannot recognise feathers, skin or hair as food anymore. Eulan is very

Fig. 4.9 Okawango diorama under construction. When the background had been painted, the animals were placed in front to find out the best composition

Fig. 4.10 Lion diorama from Amboseli in Kenya

effective, much more effective than the very dangerous arsenic poison used in former times.

Plant material like wood, moss and roots should best be deep frozen in a freezing room for a couple of weeks, for killing insects living in these materials. Before closing the diorama, insect poison based on Pyrethrine should be sprayed on all the materials in the diorama in order to kill all remaining insects. Above the diorama window, I used to leave an opening, closed with a piece of plywood or similar attached with screws, so that it could easily be opened for changing the neon tubes and for spraying in some Pyrethrine once or twice a year. This way, I never had any insect problems in the dioramas. The most effective way of protecting collections against insects is to use the blue light insect traps of the same kind that you see in shops and restaurants. Just put on the light for the night and all living insects will go there. Thus, the maintenance does not include any dangerous poisons.

4.4 The Rise and Fall of Dioramas

Dioramas might also be a part of history. Personally, during my lifetime I have seen that dioramas I had made in earlier years, from then typical biotopes and animals belonging to this environment, have become rare because of changes in peoples' lives. These changes are indicated by different forest managements, drying swamps, cattle keeping and changed agriculture. These changes might not be negative, but they are changes nonetheless. The dismantling of dioramas at the end of the twentieth century onwards seems to have ceased with new dioramas being installed. The recognition of their unique role in educational and in a historic record is finally being acknowledged.

The picture book 'My Zoological Museum' that I published in recent years (Granqvist 2012), is all that is now left of the famous diorama Hall in Helsinki. Some of the pictures are shown in this chapter.

Germany and Sweden suffered similar destructions with regards to Dioramas, but on a much smaller scale. Many natural history items with historical, cultural and artistic values have been destroyed in the past. The International Council of Museums (ICOM) and its subcommittee, the International Committee for Museums and Collections of Natural History (NATHIST), have been trying to stop these kinds of destruction and they have been greatly successful in doing so. Many dioramas that had been chosen for removal have been restored and saved. "The original collectors, conservators and taxidermists are part of the history of the specimens and thus part of the Intangible Heritage of the museum and its collections." (Norris and Granqvist 2006, p. 11).

The ICOM code of ethics states: "Museums preserve, interpret and promote the natural and cultural inheritance of humanity. Museums are responsible for the tangible, intangible and natural and cultural heritage. Governing bodies and those concerned with the strategic direction and oversight of museums, have a primary responsibility to protect and promote this heritage as well as the human, physical and financial resources made available for that purpose" (ICOM 2013).

4.5 Conclusion

Dioramas are an appropriate instrument when it comes to represent animals in their natural habitats. Therefore, they are a very important part of museum collections and should be created and maintained. Not only do dioramas show the present, but they can also be seen as windows to the past for animals and habitats long gone. In this case, dioramas themselves are sometimes artefacts of an ever-changing world.

Taxidermy, the act of preserving animals, is an art form involving craftsmanship, but also knowledge about the animal itself and the environment it normally lives in. Therefore, when crafting a diorama, the taxidermist has a lot to consider: Technical aspects, such as the correct size of the diorama, the lighting, the background and the scenery have to be considered. But the viewers' perspective plays an important role as well, when trying to arrange an authentic diorama. Collecting the materials for a diorama can be a long process, which often involves a lot of work. For good and authentic representations, it is often necessary for the creator of a diorama to visit specific locations and to acquire knowledge, perhaps even to hunt appropriate specimens.

Just as important as its creation process is the preservation for future generations. Inexperienced museum staff is one of the biggest issues and can lead to the dioramas' destruction. Teaching preservation techniques is a crucial part to save dioramas for future generations. Organisations, councils and committees, such as ICOM/NATHIST, know about the educational, cultural and artistic value of dioramas. Therefore, they are doing their best to preserve them and have been quite successful in recent years.

References

Granqvist, E. (2012). *My zoological museum*. Estonia: Greif Printhouse.

ICOM. (2013). *ICOM Code of ethics for natural history museums*. http://icom.museum/uploads/media/nathcode_ethics_en.pdf

Norris, A., & Granqvist, E. (2006). Case studies – Taxidermy and habitat groups. *ICOM Natural History Committee Newsletter, 17*, 1–12.

Eirik Granqvist is a taxidermist specialised in exhibition techniques, reconstruction of extinct animals and humans, diorama-constructions, greater mammal taxidermy and conservation of natural history objects. He has worked in many different museums (e.g. Musée des Sciences Naturelles in Orléans, France; Zoological Museum of the University of Bergen, Norway; Zoological Museum of the University of Helsinki; Museum of Natural History of Borgå, Finland) and is the founder and former Director of Préhistorama, the Museum of Human evolution and the Life in Rousson, France. Eirik is also a honorary staff member for life-time at Shanghai Science & Technology Museum, China. He gives lectures in natural history museum techniques world-wide. Eirik has founded taxidermist education schools in China and Thailand and has received numerous awards and nominations for his work.

Chapter 5
Dioramas of Marine Bird Colonies: History, Design, and Educational Importance

Rainer Hutterer and Till Töpfer

5.1 Introduction

Birds played a crucial role in the development of museum displays and their most sophisticated examples, the habitat dioramas (Wonders 1993a, b). Breeding colonies of marine birds were among the first dioramas designed by the Swedish pioneers Gustaf Kolthoff (1845–1913) and Bruno Liljefors (1860–1939), the first being a gifted taxidermist and designer, the second a wildlife painter (Hill 1987; Wonders 1993b). Their famous displays in Stockholm and Uppsala served as role models for similar presentations in museums around the world (Kamcke and Hutterer 2015). At about the same time, Frank M. Chapman (1854–1945) created bird rock groups for the American Museum of Natural History, New York, the first in 1898, the second in 1939 (Wonders 1993b; Quinn 2006). The Field Museum Chicago built a beautiful diorama on Bering Sea birds in 1914 (Metzler 2007), the Gothenburg Natural History Museum a seabird colony in 1923 (Wonders 1993c), and many other museums followed. In Germany, the Museum of Natural History at Berlin opened a diorama of a bird colony on Iceland in 1918 (Köstering 2015), and already in 1912, Alexander Koenig (1858–1940) had started to create a bird rock diorama in his new museum building in Bonn, which was finally opened in 1934 (Bechtle 1978; Hutterer 2015). In the following, we review the history and construction of this bird rock diorama in the Museum Koenig and also analyse its design from an educational point of view. We will show that a diorama does not only offer an opportunity for visitors to 'stand and stare' (Tunnicliffe and Scheersoi 2015) but also may provide a multitude of historical narratives on people who built it as well as on expeditions, science, and art.

R. Hutterer (✉) · T. Töpfer
Zoologisches Forschungsmuseum Alexander Koenig, Bonn, Germany
e-mail: R.Hutterer@leibniz-zfmk.de

A. Scheersoi, S. D. Tunnicliffe (eds.), *Natural History Dioramas – Traditional Exhibits for Current Educational Themes*,
https://doi.org/10.1007/978-3-030-00175-9_5

5.2 Case Study: Bird Rock on Bear Island, Bonn (1912–1933)

The bird rock diorama is found on the first floor of the museum. It was part of an ensemble of originally eleven large habitat dioramas (in the sense of Wonders 1993b) constructed for the Museum Alexander Koenig from 1912 to 1933, nine of which persist until today. Although planning and collecting of various components of the diorama occurred much earlier, its set-up and finishing took place between 1927 and 1933. Originally it was accompanied by two other large dioramas on African and on Palaearctic vultures, both of which were dismantled in 2000 (Hutterer 2014).

No written documents, plans, drafts, or models of the diorama have been left by the taxidermists and scientists involved. We therefore scrutinized the diorama itself, and examined specimens in the scientific collection and their labels, studied catalogues of the bird collection, scientific reports and notebooks of expedition members, and other archival material (Appendix). We also received first-hand information from the museum's retired chief taxidermist Wolfgang Hartwig who was involved in the various post-war renovations of the dioramas (Hutterer et al. 2016).

Expeditions and Scientific Background The diorama shows the birdlife on a rocky cliff of Bear Island. Why did Alexander Koenig choose this topic for his new museum? The answer is that in 1910 Koenig was already running a private zoological research institute, to which he planned to add a large building with further research facilities and public displays. The plan was that the displays would mirror, at least in part, his own research activities in Africa and Europe (Hutterer 2014). Fieldwork on Bear Island and other groundwork for the diorama were already performed since 1905.

Bear Island is the southernmost island of the Norwegian Svalbard archipelago, about halfway between Spitsbergen and the Norwegian North Cape. It has a triangular shape and covers about 178 km^2; its northern part is a lowland plain crossed by lakes and streams, while the southern tip and the southwest of the island are mountainous. The steep cliffs of the southern coast and of some offshore islets, such as the famous Stappen, are places where large bird colonies (Fig. 5.1) are found (Swenander 1900; Koenig and le Roi 1911). Alexander Koenig visited the archipelago for the first time in 1905 and subsequently organized two expeditions (in 1907 and in 1908; Koenig and le Roi 1911), during which all data and material for the diorama were assembled. At that time these islands were still nobody's land but already in the focus of commercial activities by various nations, including Germany (Lüdecke and Brunner 1912). After First World War, Norway was given sovereignty over the islands in the Svalbard Treaty of 9 February 1920 (Central Intelligence Agency 2016).

Koenig was fascinated by the fact that the bird fauna of the Spitsbergen archipelago was still fairly unexplored (Koenig 1938; Stanislaw-Kemenah 2012), although he was not the first ornithologist to work there. Gustaf Kolthoff, among

Fig. 5.1 Sea bird colony on the southern rocks of Bear Island, photographed during Alexander Koenig's expedition to Spitsbergen by Robert Fendler in June 1908 (ZFMK archives)

others, visited Bear Island in 1898 and Gustaf Swenander (1874–1943) did so in 1899. Both ornithologists subsequently published accounts on the local marine birdlife (Swenander 1900; Kolthoff 1901). Koenig, who knew Kolthoff from his visits to Sweden and Norway, may have been stimulated by their work and decided to explore these islands himself. In July 1905 Alexander Koenig, his wife Margarethe (1865–1943; Hutterer 2011), and an unnamed taxidermist entered a tourist vessel in the port of Hamburg. The voyage led to Norway and further on to Bear Island and Spitsbergen. Although it was merely a tourist cruise, Koenig had asked for the permission to collect birds and eggs (Koenig and le Roi 1911).

In 1907 Koenig hired the Norwegian vessel "Eric Jarl" which served as basis for him and his crew of more than 30 people. He was accompanied by his young assistant Dr. Otto le Roi (1878–1916) and his friend and ornithologist Hans Geyr von Schweppenburg (1884–1963; Fig. 5.2), his schoolmate Ernst Nolda, lawyer Josef Waeles, medical doctor Dr. Esau, the St. Petersburg-born painter Johannes (also Hans or Iwan) Schultze (c. 1874–1937), his chief taxidermist and photographer Robert Fendler sen. (1881–1918), and further taxidermists, workmen and kitchen maids. In 1908 Koenig hired the German vessel "Strauss" which left from Bremen via Trondheim to Tromsø; from there the voyage led again to Bear Island and Spitsbergen. Again, O. le Roi, H. Geyr von Schweppenburg and J. Schultze accompanied Koenig. The personal notebook of le Roi also lists the aeronautics expert Karl Essingh (Bonn), Hans Freiherr von Berlepsch (1857–1933) and the medic Friedrich Lentz (Stettin) as part of the scientific staff. Technical staff included the famous gunsmith Johann J. Reeb (Bonn; 1843–1924) and a group of five taxidermists led by Robert Fendler sen. that included Mr. Stumpf (Plauen), Mr. Petzold (Leipzig), Willy Vornefeld (Münster), and Hermann Sass (Anklam).

Fig. 5.2 Alexander Koenig's scientific companions Otto le Roi (left), Hans Geyr von Schweppenburg (right), and a local assistant climbing the bird cliffs on Bear Island in 1907 (ZFMK archives)

The scientific output of the two expeditions was published a few years later in a deluxe monograph entitled "Avifauna Spitzbergensis" by Koenig and le Roi (1911). In part I Koenig describes the itinerary, geography, fauna and flora. The work is superbly illustrated with photographs (by R. Fendler sen.) and with colour plates of birds and eggs painted by Archibald Thorburne (1860–1935), John Gerrard Keulemans (1842–1912), Johannes Schultze and G. Krause (no data). The second part was edited by le Roi and includes the scientific results. He gives account on the avifauna of Bear Island and Spitsbergen and also compiles an extensive ornithological bibliography on the area. Also included are further accounts on Coleoptera (beetles), Diptera (flies), Hymenoptera (sawflies, wasps, bees, ants), Trichoptera (caddisflies), Aphaniptera (fleas), Araneae (spiders), Tardigrada (tardigrades), mosses and higher plants, authored by specialists in the respective fields. Five new species were described therein. Twenty years later, Salomonsen (1932) described the northern population of the Common Murre as a new subspecies, *Uria aalge hyperborea*, and based his description mainly on Koenig's specimens from Bear Island and Spitsbergen. The holotype and paratypes of this form are kept in the Museum Koenig, and some of them are apparently in the diorama (van den Elzen 2010).

Setting The bird rock diorama has a surface area of about 28 m^2, and an inner height of 5 m. The alcove is formed by brickwork walls laid in 1912. The diorama is visible from the front through a plain glass window of 5.4 × 1.9 m. An opaque operational window of 0.7 × 2.0 m is present on the left side of the glass window.

Artificial light comes from behind the front wall. Rear, side walls and ceiling are lined with a curved wire mesh and plaster construction that obscures all edges and supports the background painting. Foreground and left and right sides are dominated by artificial rocks. The rocks are supported by a rough construction made of wood laths, on which the rocks were modelled using chicken wire, plaster, cotton tape, and paint. There are differences in appearance and colour between the rocks on the left side and in the foreground and those on the right side. The steeper cliff of the right side consists of more rectangular rocks of darker brownish colour, while the rocks of the left and centre are more rounded and more greyish-brown. Repetitive details of the cliff's right part indicate that plaster casts from real rocks were made and used several times. Whether moulds were made directly on Bear Island in 1907/08 or later somewhere in Germany is not known. However, the shape of the rocks seen on photographs of the Bear Island cliffs (Figs. 5.1 and 5.2) is similar to that of the artificial cliff in the diorama. The landscape is completed by a number of real boulders in the foreground. On the left side, in the Kittiwake colony, a few patches of moss and rotten grass brought in by the birds for nesting are found. The background painting forms an essential part of the diorama and is described below.

Originally the diorama was protected across its whole front by an awning-like construction of dark green canvas in order to avoid reflections on its window. Left and right sides had removable curtains. Once inside, the visitor could push a button that would switch on the light of the diorama for a short interval of 3–5 min (W. Hartwig, personal communication, 2016). The curtain was removed in the 1960s (Eisentraut 1984), and permanent light installed. It is possible that further details were modified subsequently, but in principal the diorama is in its original condition. In the 1980s, an electronic system was installed with a tablet in front of the diorama, from where visitors could select a certain bird species by pushing a button which would aim a spotlight at the respective specimen. Some years later the electronics failed to work and the system was dismantled. During a general renovation of the museum building from 1999 to 2003 the diorama became heavily dusted and partly damaged, but a thorough restoration by Hartmut Schmiese and the museum's taxidermy staff in 2006 restored the original condition.

The Birds The diorama contains an overwhelming diversity of birds not easy to distinguish at first sight (Fig. 5.3a), comprising nine bird species of three bird families. In total, there are 98 mounted birds on the rocks and cliffs, of which 33 are Common Murres *Uria aalge* with four nestlings, 22 Thick-billed Murres *Uria lomvia*, one Razorbill *Alca torda*, five Atlantic Puffins *Fratercula arctica*, six Black Guillemots *Cepphus grylle*, and three Little Auks *Alle alle*; while two are Glaucous Gulls *Larus hyperboreus*, 16 Black-legged Kittiwakes *Rissa tridactyla* with two nestlings, and four Northern Fulmars *Fulmarus glacialis*. In addition, there are 17 murre eggs in different conditions. The former six species belong to the family of auks (Alcidae), whereas the two gulls are members of the gull family (Laridae) and the fulmar is the sole representative of the otherwise speciose tubenoses (Procellariidae).

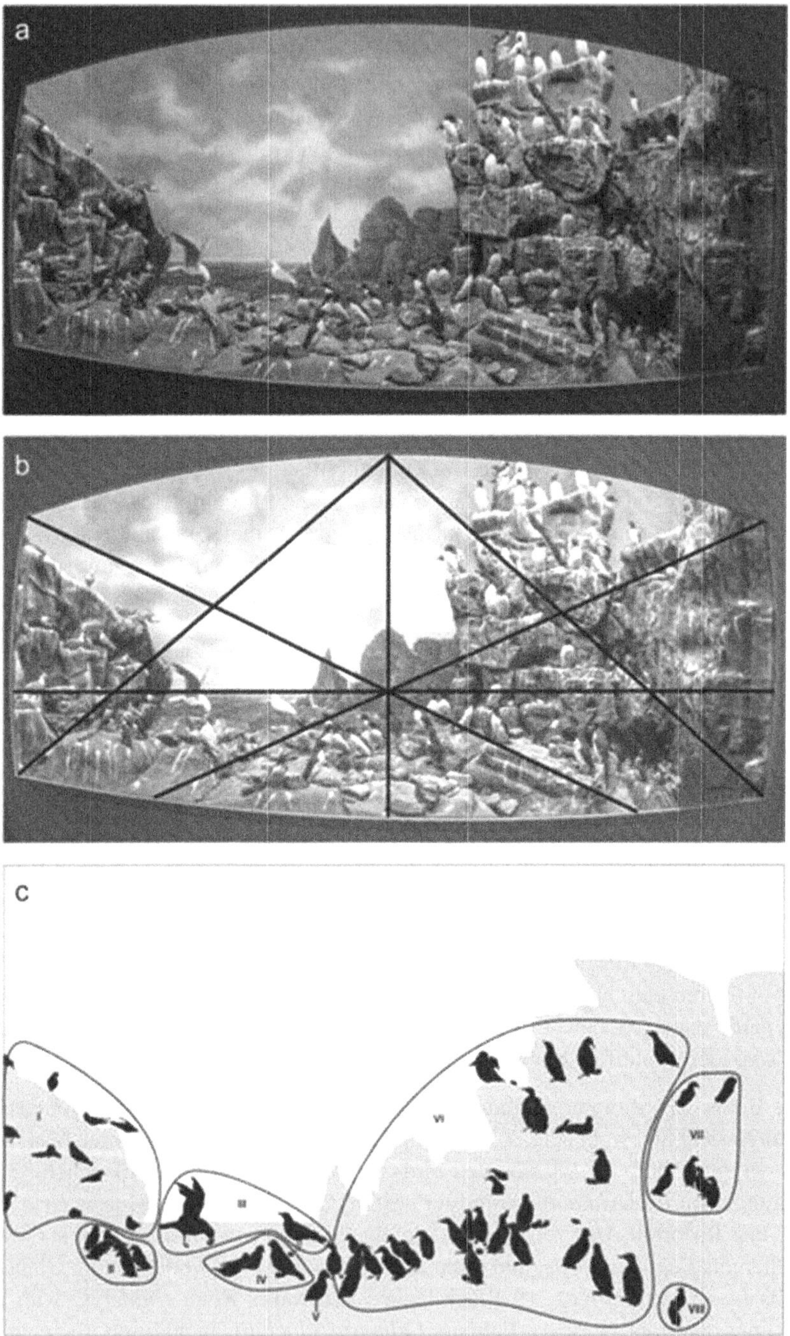

Fig. 5.3 (**a**) Panoramic view of the diorama "Bird rock on Bear Island" at Museum Koenig, Bonn, present condition (photograph by J. Adrian); (**b**) Although being asymmetrical, the diorama was constructed applying basic principles of art design: horizontal line, diagonals, triangular, and golden section perspective; (**c**) Schematic sketch showing the positioning of the bird specimens: I – Black-legged Kittiwakes; II – Atlantic Puffins; III – Glaucous Gulls; IV – Northern Fulmars; V – Razorbill; VI – Thick-billed and Common Murres; VII – Black Guillemots; VIII – Little Auks

Taxidermy A multitude of taxidermists were involved in the construction of this diorama. The bird specimens were already collected between 1905 and 1908 and skinned or mounted by taxidermists Fendler sen., Sass, Vornefeld, Petzold and Stumpf. Some of the raw skins were sent for a mounting to H. Sander and Mr. Funk (both Cologne) in 1912. The drafting of the diorama was done by Robert Fendler sen. After his death in 1918, the final work including the construction of the rocks was done by his son, Robert Fendler jun. (1911–1981), and the new chief taxidermist Berthold Korf (1893–1981), from 1927 to 1934 (Hutterer 2015), assisted by artist Victor Stoetzner-Lund (1883–1947). The quality of the bird specimens is very heterogeneous. While many of the mounts such as the murres are still excellent, others like the puffins have faded and apparently were less well-made. A postcard from about 1950 showing part of the bird rock reveals that the positions of some birds (a gull and a murre) had been changed since.

Design and Artwork Some of the early explorers of the Spitsbergen Archipelago, such as Gustav Kolthoff and Gustaf Swenander, were scientists and artists at the same time. They studied the bird fauna, wrote scientific reports, and subsequently transformed their knowledge into fine illustrations or dioramas. In the case of Koenig's expeditions and the Bonn diorama, different functions were adopted by different persons. The final results are equally convincing.

The diorama as such was designed by applying the principles of art design: horizontal line, diagonals, triangular, and golden section perspective (Metzger 1953). The horizontal line and the centre of the entire diorama are defined by the background painting alone (Fig. 5.3b). The composition and inclination of the rocks and birds partly follow diagonals and the classical triangular. The subdivision of the artificial rocks and eye-catching structures roughly follows the golden section perspective.

The background painting was realized by the artist Victor Stoetzner-Lund (Koronowski et al. 2014) who was the only staff member to leave his signature (*V. Stoetzner-Lund, Berlin, 1933*) on the lower left corner of the wall painting, hidden by plaster rocks. He was an excellent painter who made all background paintings in the museums of Bonn and Berlin (Hutterer 2015). The background painting visually merges the three-dimensional elements of the diorama with a seemingly distant landscape. It defines the horizon (Fig. 5.3a, b); below there is the fluttering sea, above a cloudy sky which covers more than two-thirds of the background. In the centre and to the right there are painted rocks which constitute an extension of the cliff casts shown in the foreground. The central part depicts the southern tip of Bear Island with the characteristic Stappen islet. The islet is pictured surrounded by a swarm of seabirds (Fig. 5.4d). On the right side, the two-dimensional background painting gradually merges into the three-dimensional bird rock, making it almost impossible for the visitor to distinguish between painted birds and real specimens as well between painted and artificial rocks. The sky is covered by dark grey clouds which open in their middle and expose a small piece of blue sky, while paler clouds form a sunlit structure in the upper centre. This creates a dramatic atmosphere, and is certainly an intended part of the composition.

Fig. 5.4 From expedition to diorama: (**a**) Historical photograph of the southern tip of Bear Island showing the striking silhouette of the Stappen islet, photographed by Robert Fendler sen. on 15 July 1907 (ZFMK archives); (**b**) Detail of an oil canvas signed 1911 by Johannes Schultze, depicting the Stappen from another perspective (original in ZFMK); (**c**) Detail of the diorama background painting, completed by V. Stötzner-Lund in 1933; (**d**) Close-up detail of the background painting, showing a flock of seabirds passing the left flank of the rock. (Photographs by R. Hutterer & T. Töpfer)

Stoetzner-Lund never saw Bear Island or another island in the North Atlantic personally. However, he did not paint the background from pure imagination. During the expeditions in 1907 and 1908, Robert Fendler took photographs of the landscape (Figs. 5.1 and 5.4a), of birds, and of many more other details; three albums with photographs are still in the archives of the museum. In addition, Fendler shot film sequences of sea and land on Bear Island and Spitsbergen, some of which show bird rocks such as the one displayed in the diorama. On both expeditions also the painter Johannes Schultze took sketches of the landscape (Koenig and le Roi 1911), and in 1909 and 1910 he executed twelve large oil paintings for Alexander Koenig, two depicting Bear Island (Fig. 5.4b) and nine Spitsbergen. In addition, A. Koenig purchased a collection of professional photographs of landscapes from the same area. All these materials were available to Stoetzner-Lund, who certainly made use of them. His background painting of Stappen islet (Fig. 5.4b) is obviously based on Fendler's photographs (Fig. 5.4a), and his bird swarm could have been based on the film sequences and photographs of Fendler as well (Fig. 5.4c, d).

5.3 Educational Potential

In educational respect, the diorama is still as impressive and informative as it was just after its opening 80 years ago. While parts of the diorama may have been carefully refurbished meanwhile, its general arrangement remained unchanged. Its greatest strength probably lies in the life-size presentation of a spectacular, yet still hardly accessible part of the world. In the figurative sense, the diorama also stands exemplarily for a habitat type whose principal structure is very similar in different bird cliffs of the northern latitudes. The Bear Island diorama thus has a dual function of reflecting something very concrete and something more conceptual at the same time. A striking facet of this diorama is that it actually bears a lot more information beyond the eye-catching scenery. Its arrangement consists of several components that illustrate different thematic aspects of a marine bird community.

Diversity of Seabirds Seabirds form an ecologically broadly similar, yet systematically heterogeneous guild of birds. They may exhibit very different lifestyles but share some general features like feeding and breeding adaptions. All marine birds invariably have to come ashore for breeding and most of the species do so in colonies, often of considerable size. These breeding aggregations frequently consist of different species that often occupy separate "layers" on a cliff where they sometimes might form mixed breeding colonies as illustrated in the Bear Island diorama (Fig. 5.3a, c).

Until today, 126 bird species have been recorded on Bear Island of which only 33 are breeding regularly on the island (Ramsar 2011). The Bear Island birdlife is estimated to comprise about 1 million seabirds during the breeding season (Ramsar 2011). Notwithstanding, the numbers of breeding pairs differ greatly between species and the predominant amount of individuals is made up by just a handful of seabird species: the most numerous species, the Black-legged Kittiwake, amounts to 130,000 breeding pairs alone (BirdLife International 2016). Reflecting this relatively species-poor, yet numerous birdlife, the Bear Island diorama at the Museum Koenig comprises only nine seabird species. The number of individuals per species shown in the diorama, however, only roughly reflects the actual ratio between the breeding birds on the island, even when taking into account Koenig and le Roi's contemporary (1911) state of knowledge who recorded merely 19 breeding bird species.

Among all seabirds, Koenig and le Roi (1911) found the Thick-billed Murre to be the most numerous species, although the Common Murre is overrepresented by one third in the diorama. The Little Auk, despite being probably the most numerous seabird species on Spitsbergen, was found by Koenig and le Roi (1911) still very abundant in Bear Island's northern part, but much scarcer on its southern shores, which might explain the only few individuals shown in the diorama which depicts the island's southern tip. Another two abundant seabird species are the Black Guillemot and the Atlantic Puffin, both shown in the diorama in a similar number even though Koenig and le Roi (1911) found the latter to be less common. In any case, the two species are distributed along the coast of Bear Island in much smaller densities than the huge breeding aggregations of murres. Finally, the Razorbill, represented by a single individual in the diorama is indeed a comparatively rare breeding species on the island (Koenig and le Roi 1911) but may itself form large colonies elsewhere (Gaston and Jones 1998). Similarly, the gulls in the diorama are mainly comprised of the colonially breeding Black-legged Kittiwakes, which are again superabundant on Bear Island, whereas there are merely two individuals of the Glaucous Gull which was found to be a common breeding species by Koenig and le Roi (1911). Finally, the ubiquitous Northern Fulmar appears in the diorama with just a few specimens although it was also found very common on Bear Island (Koenig and le Roi 1911).

The appearance of all these birds is strikingly bicoloured with dark upperparts and the underparts being white, except for the all-dark Black Guillemot (Fig. 5.5). This constellation is known as "countershading" (Ruxton et al. 2004), which is interpreted ecologically as a way to reduce the visibility of the avian body when the birds are diving and/or flying towards the respective background: a flying or swimming/diving bird is concealed against the light sky or water surface when seen from below while its darker upper side matches the darkness of the sea floor or surface when seen from above. This also explains why the deeply diving auks are predominantly blackish-brown above while the gulls and fulmars, who perform only shallow dives, have greyish upper parts with matches closely to the sea surface. Accordingly, the outer appearance of these birds can be divided roughly into two different groups: the compact, short-legged, short-winged and short-tailed auks with their spindle-shaped bodies and erect postures when at rest and the slender, long-legged, long-winged and long-tailed gulls and fulmars with their more "dynamic" physique (Fig. 5.5). The auks' characteristic body construction actually reflects a well-balanced trade-off between effective diving and flying. Although their manoeuvrability does not match the aerial skills of gulls and fulmars, they are able to reach their feeding grounds and breeding sites on the wing which allows them to flexibly react to changing food availability. Being wing-propelled divers, supported by a streamlined body shape, their short wings represent fine-tuned adaptations to maximise diving efficiency by maintaining a just sufficient flying performance at the same time. By contrast, the factually non-diving gulls and fulmars with their long wings and tails are able flyers because of lower wing-loadings and hence a much higher flight efficiency. As a consequence, auks show an intense flying style with rapid, propeller-like wing beats whereas gulls and fulmars perform a light flapping flight with extensive gliding phases.

Another interesting fact that can be discovered in the diorama is the parallel occurrence of two different plumage morphs of the Common Murre. Besides the completely black-headed morph there is bridled variant previously named "*ringvia*" which is characterised by white eye-ring that extends into a thin white, downward pointing line across the head sides, causing a "spectacled" appearance (Fig. 5.5f, f′). The relative amount of these two plumages within the different Common Murre populations varies geographically in a North-South direction. While the southern populations have no bridled individuals, their Arctic counterparts may show a 50/50 ratio (Gaston and Jones 1998). Although being the rarer morph on Bear Island

Fig. 5.5 Adult birds and eggs of the species represented in the Bear Island diorama. Specimen and egg sizes are show to scale, respectively. The specimens from the ZFMK bird collection were collected during Koenig's expeditions to Spitsbergen and Bear Island (photographs by J. Adrian). Note the two plumage variants of the Common Murre with bridled birds ("*ringvia*", f′) to appear spectacled. The two eggs for the Common Murre illustrate the high interspecific variability in shape and pattern. (**a**) Atlantic Puffin; (**b**) Black Guillemot; (**c**) Razorbill; (**d**) Little Auk; (**e**) Thick-billed Murre; (**f**) Common Murre, (**f'**) Common Murre (bridled morph "*ringvia*"); (**g**) Black-legged Kittiwake; (**h**) Glaucous Gull; (**i**) Northern Fulmar

(Koenig and le Roi 1911), in the diorama there are exhibited more bridled murres (19 vs. 14 specimens).

Breeding Ecology The steep Northern hemisphere seabird cliffs are often vertically structured (Kostrzewa 1998). Because of the specific nesting requirements and the different degree of sociability and territoriality, there usually is a layered arrangement of the different species' nesting sites which is reflected only imperfectly in the diorama, possibly in favour of visual appeal. Typically, the species that dig nesting burrows in the soil are found on top of the cliffs where the soil cover is thick enough to permit undermining. On Bear Island, the Atlantic Puffin is the only bird species that constructs such subterranean nesting sites (Table 5.1). In contrast to their breeding habits, in the diorama the puffins are arranged on the bare rock while burrows are not recognisable (Fig. 5.3a, c). On Bear Island, Little Auks sometimes re-use abandoned puffin burrows (Koenig and le Roi 1911) which results in their breeding occurrence both on top of a cliff and at its bottom, where they predominantly select breeding sites in rocky crevices or under boulders (Kostrzewa 1998). Consequently, Little Auk specimens in the diorama are shown close to a low-lying boulder area (Fig. 5.3a, c). The other two auk species that prefer breeding at similarly concealed sites are the Razorbill and the Black Guillemot, of which the latter may also occupy cliff ledges (Table 5.1). In the diorama, the position of the Black Guillemot specimens matches their natural preference for rocky crevices while the sole Razorbill is simply resting on a bare rock in the foreground (Fig. 5.3a, c). Both species, however, are not restricted to the cliff bottom and middle ranges but may instead breed along the whole flank with Black Guillemots being generally very opportunistic in the choice of their nesting sites (Gaston and Jones 1998; Kostrzewa 1998). The main surface of the cliff flanks on Bear Island is predominantly populated by dense breeding aggregations of the two murre species, forming mixed colonies as nicely illustrated in the diorama (Fig. 5.3a, c). There might be a certain tendency towards interspecific separation in some places (Gaston and Jones 1998; Kostrzewa 1998), but in general the breeding requirements of both species are very similar (Table 5.1). Only Black-legged Kittiwakes aggregate in equally large-sized cliff colonies but are spatially separated from the murres (Koenig and le Roi 1911). This is demonstrated by the "kittiwake rock" on the diorama's left-hand side (Fig. 5.3a, c). By contrast, on Bear Island the Glaucous Gull is a solitary breeder or found only in small groups with pronounced distances between individual nests (Koenig and le Roi 1911). They occupy plateaus and edges of the cliffs but do not breed within the seabird colonies as such (Table 5.1). Similarly high up in the bird rocks, but on narrow cliff ledges, Northern Fulmars breed in sometimes large colonies (Koenig and le Roi 1911). Therefore, it becomes apparent that many, but not all marine bird species are colonial breeders and that solitary breeders in one area might form substantial breeding colonies elsewhere. This different ratio between the species may result in a typical local composition of the breeding birdlife and was considered approximately in the Bear Island diorama accordingly. In doing so, the biologically correct arrangement supports the authenticity of the diorama's

Table 5.1 Some biological characteristics of the birds represented in the bird rock diorama

Species	Body length (cm)	Body mass (g)	Main food	Nest site	Egg shape	Egg size (mm)	Reference
Common Murre *Uria aalge*	38–43	945–1044	Marine fish, crustaceans	Cliff ledges to small plateaus	Conical	81.7 × 50.0	Kostrzewa (1998), Schönwetter (1963b)
Thick-billed Murre *Uria lomvia*	39–43	810–1080	Marine fish, plankton	Cliff ledges	Conical	80.2 × 50.0	Kostrzewa (1998), Schönwetter (1963b)
Razorbill *Alca torda*	37–39	524–890	Marine fish	Crevices of cliffs	Elongated-oval	75.3 × 47.6	Kostrzewa (1998), Schönwetter (1963b)
Atlantic puffin *Fratercula arctica*	26–36	460	Marine fish, plankton (crustaceans)	Burrows	Oval	63.0 × 44.0	Kostrzewa (1998), Schönwetter (1963b)
Black guillemot *Cepphus grylle*	30–32	450–550	Marine fish, crustaceans	Crevices or ledges of cliffs	Oval	58.6 × 39.3	Kostrzewa (1998), Schönwetter (1963b)
Little auk *Alle alle*	17–19	140–192	Marine plankton (crustaceans)	Rocky crevices or under large rocks	Round-oval	47.9 × 33.7	Kostrzewa (1998), Schönwetter (1963a)
Glaucous Gull *Larus hyperboreus*	64–77	1070–1820	Omnivorous	Cliff edges, rock pinnacles, slopes	Oval	76.9 × 53.7	Burger and Gochfeld (1996), Schönwetter (1963a)
Black-legged kittiwake *Rissa tridactyla*	38–40	305–512	Marine invertebrates and fish	Narrow cliff ledges	Round-oval	56.6 × 41.2	Burger and Gochfeld (1996), Schönwetter (1963a)
Northern fulmar *Fulmarus glacialis*	45–50	700–835	Marine fish, molluscs, plankton	Narrow ledges or in hollows	Oval	74.0 × 50.5	Carboneras (1992), Schönwetter (1960)

scenery and unobtrusively both communicates ornithological knowledge and underpins the sophisticated biological aspirations of the diorama's founders.

Egg Characteristics The seabirds shown in the diorama do not only differ in the position and construction of their nesting sites but also in size and in shape of their eggs. Particularly among the auks peculiar egg shapes, colours and patterns are to be found. The most striking examples are the eggs of the two murre species, being extremely conical and extraordinarily variable in colouration and patterning (Fig. 5.5e–f′). Unsurprisingly, these eggs (and the chicks) are prominent elements of the diorama that exemplify breeding and hatching as well as demonstrating failed breeding success through egg predation. The latter is illustrated by a Glaucous Gull holding a damaged murre egg in its bill, which is a regular feeding habit of these opportunistic omnivores (Table 5.1). The similarly densely breeding kittiwakes are also shown with eggs and chicks on the diorama's opposite site (Fig 5.3a, c).

Although the murres and kittiwakes (Fig. 5.5g) are the only species of which eggs are present in the diorama, it is worthwhile considering the different egg characteristics of the other species as well. One aspect is that egg sizes do not necessarily correspond to body size of the adult birds (Fig. 5.5 and Table 5.1), as exemplified by the relatively large eggs of the Little Auk (Fig. 5.5d) compared to the relatively small eggs of the large-bodied Glaucous Gull (Fig. 5.5h). On the other hand, there is a very rough correlation between the eggs' colours and patterns and the birds' breeding sites: species preferring sheltered sites such as crevices, burrows and boulders (Arctic Puffin, Little Auk) tend to have lighter and often unmarked eggs whereas species breeding on open sites tend to have darker and intensively patterned eggs (murres, gulls). This correlation is equally found in many other groups of birds, however, there are frequent exceptions. In the diorama this approximative rule is undermined by the white eggs of the open-breeding Northern Fulmar and the camouflaged eggs of the predominantly cavity-breeding Black Guillemot (Fig. 5.5i, b). The distinctive colours and patterns of murres supposedly allow for precise identification of individual eggs by the breeding parents within the huge nesting aggregations. Moreover, their extremely conical egg shape is often interpreted as a protective adaptation that prevents the eggs from falling off the narrow cliff ledges by rotating in a very small circle if jolted accidentally. Alternatively, it has been suggested that the upright stance of sitting auks requires the eggs to be conical in order to maximise body contact during breeding (Kostrzewa 1998). Both these characteristics are clearly to be seen in the diorama where a number of murre eggs allow a closer look, either next to the breeding adults or on the barren rocks in the foreground.

Feeding Ecology The presence of the different seabird species at a given location is crucially influenced by the availability of their respective food source, which overwhelmingly consists of marine animals. Differences in species composition, breeding density and breeding success can often be explained by geographically or seasonally differing food supply. Therefore, the Bear Island bird community, as illustrated in the Museum Koenig's diorama, also allows for ecological interpretations beyond the situation visualised as such. The species shown therein belong to four or five different feeding guilds, i.e. they have different preferences concerning their main food items (Table 5.1). Mainly fish-eaters are the two murre species, the Razorbill and the Black Guillemot, although the amount of other food items such as crustaceans may vary regionally or seasonally. The Northern Fulmar is also a fish-eater who flexibly complements its food with molluscs and plankton while the Atlantic Puffin regularly feeds on marine fish and plankton. Opposed to them, the Little Auk is a specialised plankton feeder. The Black-legged Kittiwake also mainly feeds on marine invertebrates but also does not spurn fish. The only true omnivorous species is the Glaucous Gull that does not only have a very broad range of food items and feeding habits (ranging from fishing over egg predation to scavenging) but is also the one species that is least bound to the sea for foraging as it will exploit food sources on the island itself. Indeed, all other species are predominantly searching for food on sea while the distances of the feeding grounds from the coast as well as diving depths (among auks) differ greatly depending on the preferred food. For example, Black Guillemots usually forage close to the coastline in shallow waters in depths below 20 m whereas murres may search for food in distances of about 30 km and will dive up to 180 m deep (Kostrzewa 1998).

Importance of Rock Islands for Bird Conservation Bear Island has become protected as a nature reserve in 2002 including a 22 km sea belt around the island (Sysselmannen på Svalbard 2005). Also, the island is a worldwide acknowledged Important Bird and Biodiversity Area (IBA SJ013) since 2014 because of its large seabird colonies and its importance as a staging area for Arctic geese (BirdLife International 2016). Moreover, since 2010 the island is listed as site no. 1966 by the Ramsar Convention on Wetlands (Ramsar 2011). All these decisions are substantially based on the existence of large seabird colonies, particularly in the island's south, and the huge numbers of bird individuals of some species, especially murres and kittiwakes. In particular, the Black Guillemot and the Razorbill are listed on the Norwegian Red List as "vulnerable" and "endangered", respectively (Ramsar 2011). This highlights the importance of such rock islands for bird conservation. Their inaccessibility and sparse human population prevented the exploitation of birds and eggs as nutritional resources that were severely impacted elsewhere. Considering the large extension of Bear Island's coastline, accounting for about 40% of the island's total size and thus being the largest local habitat type (BirdLife International 2016), it becomes apparent why the protection of these seemingly barren rocks is so important for the protection of marine birdlife. In times of global warming and ongoing shrinkage of valuable habitats, rock islands such as Bear Island are symbolic for the current endeavour to conserve the world's natural heritage. Rising sea water temperatures directly affect the breeding success of Puffins and other seabirds (Price 2013). These aspects add another educational dimension to the diorama that was surely not intended when the display was originally planned.

5.4 Concluding Remarks

We have briefly shown that the bird rock diorama provides multifaceted potential for such different biological and historical narratives like (i) expedition history, (ii) design and art of the diorama, (iii) diversity of seabirds, (iv) breeding ecology, (v) egg characteristics, (vi) feeding ecology, and (vii) importance of rock islands for bird conservation.

For each of these topics, more detailed information and narratives could be extracted. For example, the expedition history of the Bonn diorama may tell interesting stories about people. Members of the two expeditions included, apart from Alexander Koenig (Hutterer 2008), characters like Hans Geyr von Schweppenburg (famous ornithologist; see Niethammer 1964), Otto le Roi (great zoologist and Koenig's assistant who died in WW1; see Geyr von Schweppenburg 1917), Hans Freiherr von Berlepsch (founder of bird conservation; see Gebhardt 1964), but also local heroes such as Karl Essingh, a specialist in aeronautics who in 1908 navigated a captive balloon during a storm across Bonn (Milarch 1908). For many of the other persons mentioned we have no or little information. However, the history of the exploration (and exploitation) of Bear Island and the Spitsbergen Archipelago has also political and conservational aspects (Lüdecke and Brunner 2012). It is also evident from our study that the Bear Island diorama resulted from research activities of Alexander Koenig and his colleagues; Wonder's (1993c) statement that, "*The German sportsman, Alexander Koenig, established his own museum in Bonn in 1912 and had dioramas installed to display his hunting trophies*", is not correct in this point.

Our example is probably representative for many bird rock dioramas in museums around the world. A careful analysis of other dioramas will certainly reveal similar background stories, some of more general and some of more local importance. A good diorama offers an opportunity for visitors to 'stand and stare' (Tunnicliffe and Scheersoi 2015), which is a value of its own and perhaps was the intention of Koenig and other museum people of his era. However, the viewing habits of visitors, particularly of the generation adapted to digital media presentations, have changed and may require additional stimuli to get a deeper interest. Fortunately, there are many ways to improve the dioramas without violating their historical and artistic heritage. Currently, the Bonn diorama is accompanied by a general text on dioramas, a historical film clip of the 1908 expedition to Bear Island, and a plate showing birds and eggs, explaining some biological characteristics. Additional interesting offers, for example, could be an acoustic background (as in the Stuttgart Museum) with sound recordings from a bird rock colony on Bear Island, or live pictures from a remote webcam from a seabird colony somewhere in Europe. We are convinced that such a combination both of historical and current scientific information could greatly complement the diorama's attractiveness and educational value also for the next generation of visitors.

Acknowledgements We wish to thank Jörg Adrian for taking photographs, Uwe Vaartjes for contributing illustrations, Stefanie Rick for critically checking the manuscript, and Wolfgang Hartwig for sharing his personal experience with the dioramas since 1957.

Appendix: Archival Material Used for This Study

(1) Fendler, Robert (1908). Film scenes taken on Bear Island and Spitsbergen. Digitized fragments shown next to the diorama. (2) Fendler, Robert (1907, 1908). Album with photographs taken during the expeditions to Bear Island and Spitsbergen. (3) Koenig, Alexander: Album, a collection of colour prints from Spitsbergen. (4) Koenig, Alexander: Handwritten list, "Liste der Vogelbälge von den Reisen nach Spitzbergen 1907 und 1908". (5) le Roi, Otto (1907, 1908). Diaries: Spitsbergen and Bear Island. (6) Schultze, Johannes (1909–1910). 12 paintings (oil on canvas) of landscapes on Bear Islands and Spitsbergen. (7) Section of Ornithology: Entry catalogues of bird specimens. (8) Historical remnants extracted from dioramas (brush, paint, bottles, glasses, pins, cigarette packages). (9) Historical postcard collection.

References

Bechtle, W. (1978). Museum Koenig – Hohe Schule des Dioramas. *Kosmos, 11*(78), 806–813.

BirdLife International. (2016). *Important bird and biodiversity area factsheet: Bjørnøya (Bear Island)*. http://www.birdlife.org. Accessed 29 Mar 2016.

Burger, J., & Gochfeld, M. (1996). Family Laridae (Gulls). In J. del Hoyo, A. Elliott, & J. Sargatal (Eds.), *Handbook of the birds of the world* (Vol. 3, pp. 572–623). Barcelona: Lynx.

Carboneras, C. (1992). Family Procellariidae (Petrels and shearwaters). In J. del Hoyo, A. Elliott, & J. Sargatal (Eds.), *Handbook of the birds of the world* (Vol. 1, pp. 216–257). Barcelona: Lynx.

Central Intelligence Agency. (2016). *The World Factbook: Svalbarg*. https://www.cia.gov/library/publications/the-world-factbook/geos/sv.html. Accessed 10 Mar 2016.

Eisentraut, M. (1984). Entwicklung und Aufbau der Schausammlung. In G. Rheinwald (Ed.), *Die Wirbeltiersammlungen des Museums Alexander Koenig* (Bonner zoologische Monographien, 19, pp. 225–238). Bonn: Zoologisches Forschungsinstitut und Museum Alexander Koenig.

Gaston, A. J., & Jones, I. L. (1998). *The auks*. Oxford: Oxford University Press.

Gebhardt, L. (1964). *Die Ornithologen Mitteleuropas – Ein Nachschlagewerk*. Giessen: Brühlscher Verlag.

Geyr von Schweppenburg, H. (1917). Otto le Roi zur Erinnerung. *Journal für Ornithologie, 65*, 435–443.

Hill, M. (1987). *Bruno Liljefors the peerless eye*. Kingston upon Hill: The Allen Publishing Company Ltd.

Hutterer, R. (2008). Auf Zucker gebaut – Alexander Koenig und sein Museum in Bonn. *Koenigiana (Bonn), 2*, 3–8.

Hutterer, R. (2011). Von Demmin nach Bonn: Leben und Wirken von Margarethe Koenig (1865–1943). *Koenigiana (Bonn), 5*, 71–86.

Hutterer, R. (2014). Die Dioramen im Museum Koenig in Bonn: Geschichte und aktueller Stand. *Kaupia, Darmstädter Beiträge zur Naturgeschichte, 19*, 37–53.
Hutterer, R. (2015). Habitat dioramas as historical documents: A case study. In S. D. Tunnicliffe & A. Scheersoi (Eds.), *Natural history dioramas: History, construction and educational role* (pp. 23–32). Dordrecht: Springer Science+Business Media.
Hutterer, R., Schmidt-Loske, K., & Böhme, W. (2016). Meisterpräparator, Buchillustrator & Tiermaler: Wolfgang Hartwig zum 88. *Geburtstag. Koenigiana (Bonn), 10*, 41–53.
Kamcke, C., & Hutterer, R. (2015). History of dioramas. In S. D. Tunnicliffe & A. Scheersoi (Eds.), *Natural history dioramas: History, construction and educational role* (pp. 7–21). Dordrecht: Springer Science+Business Media.
Koenig, A. (1938). *Autobiographie*. Bernburg: Kunze.
Koenig, A., & le Roi, O. (1911). *Avifauna Spitzbergensis. I. Allgemeiner Teil (A. Koenig), II. Spezieller Teil (O. le Roi)*. Bonn: Privatdruck.
Kolthoff, G. (1901). *Till Spetsbergen och nordöstra Grönland ar 1900*. Stockholm: Fr. Skoglunds Förlag.
Koronowski, C., Koronowski, P., & Stoetzner-Lund, V. (2014). Der Tier- und Landschaftsmaler Victor Stoetzner-Lund (1883–1947), eine Spurensuche. *Koenigiana, 8*, 87–97.
Köstering, S. (2015). Heimatgefühle in 3-D. In A. Hermannstädter, I. Heumann, & K. Pannhorst (Eds.), *Wissensdinge. Geschichten aus dem Naturkundemuseum* (pp. 160–163). Berlin: Nicolai.
Kostrzewa, R. (1998). *Die Alken des Nordatlantiks*. Wiesbaden: AULA.
Lüdecke, C. & Brunner, K. (Eds.). (2012). Von A(ltenburg) bis Z(eppelin). Deutsche Forschung auf Spitzbergen bis 1914. 100 Jahre Expedition des Herzogs Ernst II. von Sachsen-Altenburg. *Schriftenreihe des Instituts für Geodäsie der Universität der Bundeswehr München*, 88: 118 pp.
Metzger, W. (1953). *Gesetze des Sehens. Die Lehre vom Sehen der Formen und Dinge des Raumes und der Bewegung*. Frankfurt: Waldemar Kramer.
Metzler, S. (2007). *Theatres of nature: Dioramas at the Field Museum*. Chicago: The Field Museum.
Milarch, E. (1908). Aeronautisches aus Bonn. *Illustrierte Aeronautische Mitteilungen, 12*(12), 325–329.
Niethammer, G. (1964). Hans Freiherr Geyr von Schweppenburg. *Bonner zoologische Beiträge, 15*, 115–118.
Price, M. (2013). *Climate change spells peril for puffins (updated)*. National Wildlife Federation's Blog. http://blog.nwf.org/2013/06/climate-change-spells-peril-for-puffins/. Accessed 22 Sept 2016.
Quinn, S. C. (2006). *Windows on nature: The great habitat dioramas of the American Museum of Natural History*. New York: Abrams.
Ramsar Convention on Wetlands. (2011). *Ramsar sites information service*. Bear Island. https://rsis.ramsar.org/ris/1966. Accessed 29 Mar 2016.
Ruxton, G. D., Sherratt, T. N., & Speed, M. P. (2004). *Avoiding attack. The evolutionary ecology of crypsis, warning signals & mimicry*. Oxford: Oxford University Press.
Salomonsen, F. (1932). Description of three new Guillemots (*Uria aalge*). *Ibis, 74*, 128–132.
Schönwetter, M. (1960). *Handbuch der Oologie. Band I (Nonpasseres)*, 1. Lieferung. Berlin: Akademie-Verlag.
Schönwetter, M. (1963a). *Handbuch der Oologie. Band I (Nonpasseres)*, 7. Lieferung. Berlin: Akademie-Verlag.
Schönwetter, M. (1963b). *Handbuch der Oologie. Band I (Nonpasseres)*, 8. Lieferung. Berlin: Akademie-Verlag.
Stanislaw-Kemenah, A.-K. (2012). „Versunken in Gedanken auf dem schwellenden Teppiche der Polarweide …" Alexander Koenigs zoologisch-ornithologische Forschungsreisen nach Spitzbergen 1905–1908. In C. Lüdecke & K. Brunner (Eds.), Von (A)ltenburg bis (Z)eppelin. Deutsche Forschung auf Spitzbergen bis 1914. 100 Jahre Expedition des Herzogs Ernst II. von Sachsen-Altenburg. *Schriftenreihe des Instituts für Geodäsie der Universität der Bundeswehr München, 88*, 13–21.

Swenander, G. (1900). Beiträge zur Fauna der Bären-Insel 1 Die Vögel. *Bihang till Kungliga Svenska Vetenskaps-Akademiens Handlingar,* 26, afd. IV (3), 1–50, 3 pls.
Sysselmannen på Svalbard. (2005). *Forvaltningsplan Bjørnøya Naturreservat 2005–2010.* Svalbard.
Tunnicliffe, S. D., & Scheersoi, A. (2015). Dioramas as important tools in biological education. In S. D. Tunnicliffe & A. Scheersoi (Eds.), *Natural history dioramas: History, construction and educational role* (pp. 133–144). Dordrecht: Springer Science+Business Media.
van den Elzen, R. (2010). Type specimens in the bird collections of the Zoologisches Forschungsmuseum Alexander Koenig, Bonn. *Bonn zoological Bulletin, 59,* 29–77.
Wonders, K. (1993a). Bird taxidermy and the origin of habitat dioramas. In R. G. Mazzolini (Ed.), *Non-verbal communication of science prior to 1900* (pp. 441–447). Florence: Olschki.
Wonders, K. (1993b). Habitat dioramas: Illusions of wilderness in museums of natural history. *Acta Universitatis Upsalensis, Figura, Nova Series, 25,* 1–262.
Wonders, K. (1993c). Habitat dioramas as ecological theatre. *European Review, 1,* 285–300.

Rainer Hutterer is curator emeritus in the vertebrate department at the Zoologisches Forschungsmuseum Alexander Koenig in Bonn, Germany. His main research is on systematics and evolution of mammals with emphasis on small mammals of Africa. He has also participated in archaeological projects and studied animal remains. History of science is another field of interest; he has published numerous articles on the history of the Museum Koenig, including dioramas, and on the history of the German Society of Mammalogists.

Till Töpfer works as bird curator and head of the ornithological section at the Zoologisches Forschungsmuseum Alexander Koenig in Bonn, Germany. His scientific interests comprise collection-based research and field studies on systematics, taxonomy, phylogenetics and zoogeography of birds using combined comparative methods (molecular genetics, morphology, ecology).

Chapter 6
Botanical Dioramas – Just Beautiful?

Kathrin Grotz

6.1 Introduction

The majority of dioramas on display at natural history museums throughout Europe and North America put their emphasis on animals, with artistically painted natural landscapes in the background and life-sized plants in the foreground. The latter are carefully dried and recoloured specimens, or threedimensional models elaborately made from wax, plaster or papier-maché. In many cases, only a limited number of plants has been selected, aiming to emphasize the most conspicuous floristic inventory of the respective animal's habitat (Potztal 1975). Although their reproduction is usually of high scientific and artistic quality, the role of plant models in most dioramas is limited to setting the natural "stage" for numerous stuffed animals, which thus remain the all-time favourite of visitor perception.

Rarely enough, diorama artists focus on the exclusive representation of plant environments. An outstanding example are the seven life-sized window dioramas depicting different biomes of the United States on display at Botany Hall of the Carnegie Museum of Natural History in Pittsburgh, Pennsylvania. Most of them were created by the Austrian Ottmar von Fuehrer and his wife Hanne between the 1920s and 1960s (Wonders 1993). Life-sized plant dioramas like these, however, pose some specific challenges. In order to achieve a realistic and striking impression of vegetation, a large amount of individual plants is necessary. If the diorama maker decides to work with real plants or plant parts, these must be collected, preserved in a threedimensional way, and mounted. Not all plants are suited for this process, and those passing the preservation procedures always remain potentially threatened by humidity or pests. Artificial plant models, on the other hand, might

K. Grotz (✉)
Botanic Garden and Botanical Museum, Freie Universität Berlin, Berlin, Germany
e-mail: k.grotz@bgbm.org

A. Scheersoi, S. D. Tunnicliffe (eds.), *Natural History Dioramas – Traditional Exhibits for Current Educational Themes*,
https://doi.org/10.1007/978-3-030-00175-9_6

not always appear natural. In addition, the assemblage of plant groups requires hundreds of individual models or specimens. Mass-produced, uniform plant models are not suitable if the intention is to create a "natural" visual appearance. Each single model therefore needs to be manufactured individually. All these factors make the production of life sized plant dioramas extremely challenging and time consuming (Potztal 1975).

Life-sized plant dioramas also have a limitation with regard to their content: Due to space restrictions, they cannot offer a complete overview impression of a vegetation type with its numerous different plant communities. Instead, most of the dioramas showcasing plants close up on a specific plant community (Potztal 1975). An excellent and most fascinating example is the exact reconstruction of a meadow piece, using hundreds of preserved and mounted original plant specimens, on display at the Museum of Natural History in Görlitz, Germany.

6.2 Dioramas in the Context of a "Model Museum"

If the representation of vegetation is intended to go beyound a close-up view on plant communities, diorama builders have to resort to a scale smaller than life. Karen Wonders describes these "miniature dioramas" which contain models of animals and plants, as a perfect means to communicate environmental information within a developmental series. As an example, she identifies 24 miniature dioramas at Harvard Forest Museum, which "… represent in progressive stages the history of central New England forests during more than 200 years of settlement, beginning with the forest in its primeval state, and ending with the commercial exploitation of the forest." (Wonders 1993, pp. 20). If we look at the sixteen miniature dioramas on display at the Botanical Museum in Berlin, this developmental aspect holds true only for three paleobotanic dioramas representing the developmental stages of plant life during the Devonian, Carboniferous and Mesozoic ages. The other thirteen miniature dioramas form part of a completely different conceptual scheme. Ten dioramas in scales 1:10 and 1:20 represent different vegetation types as completion to the permanent exhibition's plant geography, and three dioramas show plantations of sugar cane, cotton and cocoa, thus contributing to the economic plants section.

According to an unpublished inventory list (folder "Botanische Werte"; BGBM archive) all dioramas were built between 1956 and 1967. At this time, dioramas had been around in natural history displays worldwide for quite a while. On the other hand, many botanical museums had already ceased their exhibition activities at that point. This relatively late genesis, is owed to the fact that the Berlin Botanical Museum's permanent exhibition was developed from scratch in the postwar era between the late 1950s and 1991. The Museum building, inaugurated in 1907 and located on the premises of the Berlin Botanic Garden, had suffered severly in the Second World War. Bombing destroyed almost all herbarium and library collections and devastated more than 1200 square meters of exhibitions.

The showrooms of the old Botanical Museum were loaded with objects, most of them "originals": plant parts in alcohol, wood samples, plant products and ethnobotanical objects. Numerous illustrations and panels as well as some very few plant models provided additional information. (Potztal 1968; Grotz 2014). The exhibition conceived by Adolf Engler (1844–1930), director of the Botanic Garden and Botanical Museum, focused on the university curriculum of Botany: "It shall provide the student and every person seeking instruction [...] an overview of the most important phenomena of plant life, plant history, the distribution of plants and the use of plants". In the stern professor's imagination visitors should not seek recreation, neither in the Garden nor at the Museum, but rather be prepared for serious floristic studies (Engler 1909). The showrooms, located in the same building as the lecture halls and laboratories, provided an added learning place for lectures and practical demonstrations. Generations of students used the exhibition as an integral part of their biology studies at the Friedrich-Wilhelm-University of Berlin. At the same time, Engler's exhibition considered domestic and tropical economic plants and their products – teaching practical knowledge to aspiring colonial planters and traders (Grotz 2015a).

Museum objects that could be saved from the ruins of war were made accessible to the public again in late 1946. However, those in charge knew the substance of the "old" Botanical Museum had been irretrievably lost, and they started to study then "modern" trends in museology, pouring into Germany from the Anglo-Saxon world. Eva Potztal (1924–2000), herself trained as a botanist, conducted the rebuilding of the exhibition wing from 1959 until her retirement in 1990 with iron will and inexhaustible energy. Other botanists as well as permanent and freelance artistic staff supported her in this undertaking (Hiepko 2000).

The lack of original showpieces became the driving force for fundamental changes. "This museum shall be a center of attraction instead of a deathlike depot, it shall be educating the public in the truest sense of the word, and at the same time keep up to the highest standards in order to stimulate an attract young academics." (letter, BGBM acting director Theo Eckhardt to the Berlin Senator for Public Education, December 11, 1959; BGBM archive). In keeping with the standard botanical textbooks, the exhibition concept and content remained committed to university audiences. However, in order to make the Botanical Museum more attractive for popular audiences, exhibition design and object selection were radically altered. "Visitors shall not only learn economic and botanical facts, but shall also be aesthetically satisfied by the form of presentation" (letter, BGBM director Walter Domke to the Berlin lottery, June 22, 1961; BGBM archive). Instead of an overwhelming wealth of herbarium specimens, visitors encountered a carefully curated selection of models and small dioramas. Colouring of the surrounding exhibition environment, consistent typography and especially designed display cases contributed to a unique exhibition ensemble, which forms the core of the permanent exhibition in the Berlin Botanical Museum to this day (Grotz 2014).

6.3 The Making of Botanical Dioramas

Among the first objects constructed for the new Botanical Museum were two miniature dioramas, "Rainforest in Sumatra with giant *Rafflesia* and *Pandanus*" and "Rocky caatinga savanna in NE Brazil with *Cavanillesia*", both 1:10. In January 1956, the director of the Berlin trustee savings Bank donated the required funds of 2000 Deutsche Mark (DM), after he had been convinced of the undertaking by a homemade model prototype (letter, G. M. Schulze to BGBM acting director Theo Eckhardt, April 5, 1967; BGBM archive; Gerloff 1984). This financial starting signal subsequently helped to acquire numerous other funds by the Berlin lottery and the Berlin construction plan (Hiepko 2000). But even more important, the initial work on dioramas conducted by Heinz Woern (1914–1962), a trained grafic designer and cartograph, eventually triggered a creative big bang, inspiring the curators involved to create a museum made up entirely of models and dioramas (Potztal 1975). Woern himself had been taken by colleagues on an excursion to Salzburg. There he had visited the "Haus der Natur", and was fascinated by the lifesize habitat dioramas of Wolfgang Graßberger. Upon his return to Berlin, he set out for his first own miniature diorama of a tropical rainforest, which he completed within a year. In early 1959, Woern was entrusted the artistic and technical lead of the museum undertaking (Potztal 1960). Until his sudden death in 1962, with the help of up to six technical and artistic staff, he constructed three dioramas for the paleobotany section: "Devonian landscape", "Landscape in the Age of coal", "Mesozoic landscape", all 1:20. Also, Woern and his staff completeted several boxed dioramas for the economic plants section: "cotton plantation", "sugar plantation" and "cocoa plantation", all 1:20 (Domke 1964).

The artistic and technical standards of Heinz Woern and chief curator Eva Potztal were high, as they were prepared to break new ground in the design of botanical miniature dioramas. Visits to other museums with diorama displays proved unsatisfactory, as this report by a colleague botanist from a visit to Museum Alexander Koenig in Bonn indicates: "The dioramas they have, cannot be compared to the botanical dioramas we want to create. In my opinion, some large dioramas can be addressed as dioramas in our sense of the term. With miniature dioramas, this seems questionable to me; here the animal or animal group stands before a merely adumbrated background" (G. M. Schulze, report of a visit to the Alexander Koenig Museum in Bonn, January 5 1962; BGBM archive).

The dioramas of the Botanical Museum have all been constructed according to the alcove design (Wonders 1993). "A small viewing window allows a glimpse into the respective alcove or box construction, the perspective is forced on an ascending ground plate passing over in painted, round horizons. Ten small dioramas, measuring 75 cm in width and 45 cm in maximum depth have been placed in boxes. Their ceilings consist of enveloping plaster domes, to enhance dimensional illusion. Six large dioramas with a width of 180 cm and a maximum depth of 75 cm have been built in compact alcoves underneath the galleries of the museum entrance, making enveloping domes unnecessary. The ground plate of all dioramas ascends by five

Fig. 6.1 Samoyed [Nenet] tundra between White Sea and Ural Mountains, 1:10. (Photo: C. Hillmann-Huber, BGBM)

degrees, it consists of wood or of toned glass in the case of water habitats. Landscape contours were created on the basis of a skeleton of cardboard strips, encased with glue saturated gauze, postformed with modeling clay and subsequently painted" (Potztal 1975).

Tiny plant models were then placed on the premodeled bases. Every single component of these miniatures was meticulously made by hand, partly under the magnifying glass. On one hand, uniformity should be avoided in order to achieve the most "natural" effect with plants that formed groups. On the other hand, forced perspective made custom sizing of each single plant model necessary. The preferred building materials were copper and plastic sheets of different thicknesses, very fine wires, up to 0.35 mm in diameter, acrylic glass, chenille, fibers, balsa wood, plaster and putty. Natural materials such as branches or leaves, or the prickly capsules of chestnut seeds were used in the construction of the six large dioramas. However, they proved to be not as durable in a long-term perspective, so that in the small dioramas most plant models were made of artificial materials and painted with artist's oil paints. The three alder trees standing at the "Lake of the North German lowland" 1:10 required 12,000 custom-made miniature leaves of painted copper sheet, while the dwarf birches adorning the "Samoyed [Nenet] tundra between White Sea and Ural Mountains" 1:10 consumed about 20,000 sheets turned into leaves (Fig. 6.1). With its 1.5 cm tall gentians and numerous other miniature mountain flowers the "Tree line in the Limestone Alps" 1:10 demanded special patience and steady hands of cooperating sculptors, graphic artists, craftsmen and technicians. If one adds the "Atlantic coast rock formation at low tide" 1:10, the production of these four dioramas mentioned in this paragraph alone, required 18 months of intensive and concentrated work by five expert employees (Potztal 1968, 1983).

6.4 Types and Tableaux

Most of the dioramas on display at the Berlin Botanical Museum, ten out of a total of sixteen, were designed as part of the plant geography section of the permanent exhibition. This section was conceived by Fritz Mattick (1901–1984), a renowned expert in the field of plant geography (Diels and Mattick 1958; Mattick and Ern 1985), who worked as a custodian at the Botanical Museum until 1966 (Gerloff 1984). Mattick's goal was to present an overview of all vegetation types worldwide. Vegetation types are the visible result of all environmental factors at a particular location. They set the standard for the vegetational potential of a certain area. In the warmer areas it is the amount of precipitation, in cold zones the temperature that decides which vegetation type will eventually dominante the landscape (Diels and Mattick 1958).

For the exhibition, Mattick distinguished (apart from arable land) a total of 12 vegetation types: (1) evergreen rainforests; (2) savannahs and monsoon forests; (3) xeric shrublands; (4) mangroves and swamp forests; (5) high mountain formations; (6) semi-deserts and deserts; (7) sclerophyllic vegetation; (8) steppes; (9) deciduous forests; (10) coniferous forests; (11) tundra; (12) fresh water and salt water plant communities.

In order to show the global distribution of these vegetation types, Mattick designed a huge half-relief wall map on a scale of 1:80 Mio. In a second step, he intended to highlight the many different manifestations of these idealized vegetation types in the natural landscapes worldwide. For this purpose, Mattick had luminous dots placed on the large wall map. By pressing a button, both the dot on the map as well as a large photographic slide attached to the side of the map lit up, thus linking the region marked by the light dot with a picture of the specific vegetation found there. By comparing the viewer should learn that, for example, vegetation type 1: evergreen rainforests would look different in Africa and in South America due to different species inventories.

Similar to the large slides, although spatially much more distributed in the museum, the dioramas too presented "vegetation tableaux" which could easily be attributed to distinct geographic localities by their characteristic plant species inventory (Potztal 1968). In this sense, the miniature dioramas of the plant geography section are similar to what Karen Wonders calls "zonal group", taking "as their theme the influence of altitude and climatic zones on the evolution and distribution of animals" (Wonders 1993, 21).

In contrast to early museum concepts, which optimistically envisaged 35 miniature dioramas (note by G.M. Schulze, undated [9/1957?]; BGBM archive), by 1967 a total of ten dioramas represented seven local manifestations of the 12 vegetation types enlisted above (Mattick and Ern 1985):

- "Rainforest in Sumatra with giant *Rafflesia* and *Pandanus*" 1:10 (1: evergreen rainforests)
- "Savannah in east Africa with *Adansonia*" 1:20 (2: savannahs)

Fig. 6.2 Tree line in the Kenia mountains (East Africa) with *Lobelia* and *Senecio*, 1:20. (Photo: C. Hillmann-Huber, BGBM)

- "Rocky caatinga savanna in NE Brazil with *Cavanillesia*" 1:10 (3: xeric shrublands)
- "Mangrove vegetation of the West African coast" 1:20 (4: mangroves)
- "Swamp Forests with *Taxodium distychum* in the coastal areas of South Florida" 1:20 (4: swamp forests)
- "Tree line in the Limestone Alps" 1:10 (5: high mountain formations)
- "Tree line in the Kenia mountains (East Africa) with *Lobelia* and *Senecio*" 1:20 (5: high mountain formations) (Figs. 6.2 and 6.3)
- "Samoyed [Nenet] tundra between White Sea and Ural Mountains" 1:10 (11: tundra)
- "Lake of the North German lowland", 1:10 (12: fresh water communities)
- "Atlantic coast rock formation at low tide" 1:10 (12: salt water communities)

The vegetation dioramas address a double message to the visitors: on the one hand, they represent abstract vegetation types which, given the right climate and/or altitude, could occur anywhere in the world. On the other hand, each of these "vegetation tableaux" show a very specific region of the world, where this vegetation type is characterized by very distinct plant species (character plants), and the viewer shall recognize this. Besides the plants, attributed figurines, such as a woman in a traditional Samoyed [Nenet] costume in the tundra showcase or buildings like the Bavarian hayrick in the alpine scenery support this curatorial goal of local differentiation.

Fig. 6.3 Tree line in the Kenia mountains (East Africa) with *Lobelia* and *Senecio*. (Photo by Robert E. Fries before 1923)

But how "realistic" are these tableaux? Do they depict real places as the photographic slides besides the large vegetation map? Eva Potztal was in favour of composing artficial sites. She writes that an idealized vegetation type has to be constructed from several individual images which she would pick from current botanical literature or from slide series images, while staying as true to life as possible in (plant) details, (Potztal 1975). What this exactly means and how the curators implemented this in practice has not been preserved in the archives of the Botanical Museum. However, a quick check of a standard work at the time (Schimper and Faber 1935, tab 500), and the author found a photograph of the west side of the Kenya-volcano at 3500 m above sea level, with a striking visual similarity to the diorama "Tree line in the Kenya mountains (East Africa) with Lobelia and Senecio" 1:20. It thus seems that real landscape situations could also have served as models to the vegetation dioramas, but this still needs considerable research.

6.5 Too Subtle to Succeed?

The miniature dioramas are among the treasures of the Botanical Museum. For children and adults, they are equally fascinating because of their richness of detail. Since the renovation of the Botanical Museum in 2005/2006 (Grotz et al. 2006), visitors can take an even closer look at the dioramas. Instead of guard railings, there are now stages that slide out and allow for even the little ones, to take a good look at the different settings.

While the aesthetic and emotional value of the dioramas is undisputed, most visitors to the Botanical Museum find it difficult to understand what they convey. The complex two-stage exhibition concept involving vegetation types as well as their local manifestations, needs a selective tour of the complete exhibition space, since many of the dioramas are located far away from the giant plant geography map which serves as a centerpiece for explanations. In the time of Mattick, the majority of museum visitors came as groups on guided tours, mostly school classes or university courses. The few individual visitors were usually well informed or willing to read background information. Today, the museum has opened to a broader audience with much wider-ranging interests and a more diverse educational background, receiving far more individual visitors than groups.

Furthermore, the visitor's perception has changed significantly over the past decades. Wheras these miniature worlds in the 1960s for West Berlin visitors of the Botanic Garden and Botanical Museum might have been a small escape from everyday life in the shadow of the Berlin Wall, many of the vegetation types are known today by the vistor's own experience and travels. Given the changes in viewing habits and compared to multifold media options, a "peep in the box" appears out of fashion (Kutner 2015).

Considering this, there is a third issue: we need to establish a stronger tie between vegetation images and today's reality, which is characterized by ever more rapid climatic and ecological change. Although it was definitely Fritz Mattick's intention to sensitize the general audiences for the individual's responsibility towards nature (Gerloff 1984), his approach to achieve this awareness through the mediation of basic knowledge about flora and vegetation of the earth today seems overly academic and far too complex.

With this in mind, the Botanical Museum started a colaboration with students of illustration at the Berlin University of Arts in the winter semester 2014/15. Four selected dioramas were analyzed and then set up against an artistically created "update" from today's perspective. Among the projects was also an examination of the rainforest, the oldest diorama of the collection. The students came up with a short animated visualisation in the form of a slowly disintegrating leaf. This simple image reminds viewers in a quite touching way that in the 60 years since the creation of the "rainforest" diorama, more than half of the forests in the Malay Archipelago have been irretrievably lost and that this trend continues (Grotz 2015b). Perhaps this approach could be one way to bridge the gap between idealized vegetation types and the realities of threatened habitats? The emotional and aesthetic force of these miniature treasures, grasp the visitor's attention by creating add ons that make use of a deliberately artistic language, instead of an enumeration of scientific facts.

References

Diels, L., & Mattick, F. (1958). *Pflanzengeographie* (5th ed.). Berlin: De Gruyter.

Domke, W. (1964). Fritz Woern: 1914 bis 1962. *Willdenowia, 3*, 653–656.

Engler, A. (1909). Botanische Museen und deren Aufgaben mit bes. Berücksichtigung des neuen Botanischen Museums in Dahlem bei Berlin. *Internationale Wochenschrift für Wissenschaft, Kunst und Technik, 3*, 72–84.

Gerloff, J. (1984). Prof. Dr. Fritz Mattick *(17.5.1901 – 3.1.1984)*. *Willdenowia, 14*, 3–13.

Grotz, K. (2014). „Das Flair der echten Pflanzen". Botanische Modelle und Kleindioramen im Botanischen Museum Berlin-Dahlem. In D. Ludwig, et al. (Eds.), *Das materielle Modell. Objektgeschichte aus der wissenschaftklichen Praxis* (pp. 171–178). Paderborn.

Grotz, K. (2015a) "The devil is in the details" model-making and museum communication atz the botanical museum berlin. In K. Grotz (Ed.), *modelSHOW exhibition catalogue* (pp. 73–75). Berlin.

Grotz, K. (Ed.). (2015b). *modelSHOW exhibition catalogue* (pp. 124–127). Berlin.

Grotz, K., Lack, H. W., et al. (2006). Die Quadratur des Kreises – Denkmalschutz und Museumserneuerung. *Zur Wiedereröffnung der ersten Etage der Dauerausstellung*. Museums J. (Berlin), 1, 18–20.

Hiepko, P. (2000). Eva Potztal (1924–2000). *Willdenowia, 30*, 387–393.

Kutner, M. (2015). Museum dioramas are as endangered as the animals they contain. *Newsweek*, 165.6 (Aug 14, 2015).

Mattick, F., & Ern, H. (Eds.). (1985). *Pflanzengeographie einschließlich Geschichte der Pflanzen*. Berlin: Verl. Botanisches Museum.

Potztal, E. (1960). Bericht über den Botanischen Garten und das Botanische Museum in Berlin-Dahlem für die Jahre 1958 und 1959. *Willdenowia, 2*, 427–455.

Potztal, E. (1968). Das einzige botanische Schaumuseum in Deutschland. *Mitt.-Blatt Landesverb. Gartenbau Landw. Berlin, 19*, 39–40.

Potztal, E. (1975). Neue Ausstellungstechniken im Botanischen Museum Berlin-Dahlem. *Museologia, 4*(VI), 3–10.

Potztal, E. (1983). Der Beitrag der Botanik in der musealen Präsentation – dargestellt am Beispiel des Botanischen Museums in Berlin. In W. Kolbe & A. Biermann (Eds.), *Ökologische Aspekte im Museum* (pp. 100–104). Köln: Rheinland-Verlag.

Schimper, A. W. F., & von Faber, F. C. (Eds.). (1935). *Pflanzengeographie auf physiologischer Grundlage vol 2* (3rd ed.). Jena.

Wonders, K. (1993). *Habitat dioramas: Illusions of wilderness in museums of natural history*. Uppsala.

Kathrin Grotz, M.A., was born in Erlangen; she studied History, Political Science and Cultural Anthropology at Heidelberg University and University of New Mexico. After a scientific traineeship at Reiss-Museum, Mannheim, she began in 2002 as an outreach specialist at the Botanical Museum Berlin, since 2010 she heads the exhibition and museum section of the Botanic Garden and Botanical Museum in Berlin, managing both indoor and outdoor exhibition projects of the institution, which forms part of the Freie Universität Berlin. One of her main research interests are material models in science with a focus on botanical models; this led her to curate the exhibition modelSHOW in 2015.

Part II
Theoretical Aspects of Learning with Dioramas

Chapter 7
Dioramas as (Scientific) Models in Natural History Museums

Alexandra Moormann and Charlène Bélanger

7.1 Introduction

Dioramas appeared in the second half of the nineteenth century in natural history museums, first in Sweden and in the USA before getting adopted over a short period of time by most of the natural history museums all over the world (Bitgood 1996). One important goal put forth for this first generation of dioramas was to convey the scientific collections to a broader audience (Köstering 2015). Many dioramas, like the ones found at the Museum für Naturkunde Berlin in Germany, were built between 1918 and 1925 when most people did not have the opportunity to travel and see the natural beauty of remote places (Damaschun et al. 2010).

Nowadays, after suffering a short decline in popularity, dioramas are again receiving attention. In a time where natural history museums are positioning themselves as important players in the field of environmental education, they are considered to be powerful educational means for the promotion of natural conservation and love of nature (Ash 2004; Borg 2015; Girault and Sauvé 2008).

Typically, museum dioramas exhibit groups of taxidermized animals together with certain elements of their natural habitat (rocks, trees, grass, etc.) placed in front of a curved painted background scenery (Ash 2004; Bitgood 1996; Köstering 2015; Montpetit 1996). They are meant to provide visitors with an occasion to observe specific animal behavior, by showing one selected moment, like a picture capturing a fixed moment in time. These still depictions of nature allow visitors to take time to observe and inquire about the elements interacting in natural habitats

A. Moormann (✉)
Museum für Naturkunde, Berlin, Germany
e-mail: alexandra.moormann@mfn-berlin.de

C. Bélanger
Université du Québec à Montréal, Quebec, Canada
e-mail: charlene.belanger@mccord-stewart.ca

A. Scheersoi, S. D. Tunnicliffe (eds.), *Natural History Dioramas – Traditional Exhibits for Current Educational Themes*,
https://doi.org/10.1007/978-3-030-00175-9_7

(Van Praët 1989). Dioramas facilitate the learning about biological phenomena by presenting interacting elements together in a context (Scheersoi 2014). In this regard, one of the first champions of dioramas, the taxidermist Phillip Leopold Martin (1880) wrote that "this concept [of dioramas] allows us to bring together those creatures that share a set of living conditions and indeed live alongside each other." (Martin 1880, cited in Damaschun et al. 2010, p. 95).

At the time when dioramas appeared in museums, the study of animal behavior within certain ecosystems was an important scientific research topic. Natural history museums wanted to share this new perspective on the natural world to the non-expert public. In the specific case of the Museum für Naturkunde Berlin, Willy Kükenthal who was director of the museum in 1918, decided to change the taxonomically classified collection of the domestic fauna into a biological collection with a focus on animals' ways of living. Already at this time some of the animals which were included in the impressive large dioramas (Fig. 7.1) were in danger of extinction. Thus, their exhibition was meant to promote nature conservation (Köstering 2015).

Indeed, research in the field of museum studies has shown that the content and design of exhibitions in science museums have always been closely linked to the dominant scientific epistemic practices of the time (Fortin-Debart 2003; Girault 2003; Marandino et al. 2015). The appearance of museum dioramas coincided with the development of research disciplines within the biological sciences (such as zoology,

Fig. 7.1 Alpendiorama. One of the large dioramas in the exhibition of the Museum für Naturkunde Berlin. This diorama transports the visitor back to an alpine landscape more than a hundred years ago. It was then still inhabited by brown bears, bearded vultures and ibexes. Copyright Museum für Naturkunde Berlin

botany, anatomy, etc.) and the popularization of Darwin's theory of evolution, which induced a change of paradigm in the biological sciences (Girault 2003).

Recently, scientists, instead of aiming at the systematic classification of all living organisms on the planet based on the description and comparison of their physical structure, have started studying animal behavior and their environment as interactive systems, in which living organisms are interdependent to each other (Fortin-Debart 2003; Köstering 2015). This new interest for processes and interactions occurring within large natural systems - the ecosystems - was the foundation of *ecology* as a new way of understanding nature through which: "ecology establishes itself as a scientific method and imposes a new museological thematic." (Van Praët 1989, cited in Fortin-Debart 2003, p. 3).

From the first dioramas that appeared more than one hundred years ago to the ones that we find today in natural history museums, dioramas were always meant by their creators to trigger visitors learning about biological topics such as animal behavior and ecosystems (Tunnicliffe and Scheersoi 2015). In our recent research work, we have been interested in studying a special kind of learning that happens in the museum: the learning which is specifically related to models and model-based reasoning (Clement 2000). With this is mind we look at dioramas as simplified models of natural ecosystems. In fact, because they show patterns of interaction between animals and nature, they can be understood as visual representations of the scientific models they were based on (Montpetit 1996). Their purpose is thus mainly educational and they can give access to a scientific way of thinking about nature (Marandino et al. 2015).

In this paper we wish to answer two specific questions with regards to dioramas as models. And those questions are: *To what extent can dioramas be considered scientific models? And, how can dioramas promote model-based learning?*

In the first section of this text, we will discuss the intentions that underlie the design and presentation of dioramas in natural sciences museums. Then, we will define what is the purpose of models and modeling in the biological sciences in order to discuss the possible educational role of dioramas as models. In the third and last section, we will propose a new way of looking at the learning that occurs through the interaction with dioramas, when they are used as models in a model-based learning approach.

7.2 The Purposes of Museum Dioramas

As stated earlier, before the appearance of dioramas and other similar types of thematic exhibitions, natural science museums were mainly concerned with the exhibition of objects and specimens from their collection (Marandino et al. 2015; Köstering 2003). They were exhibiting hundreds of items behind glass, organized according to a certain logical structure, that was dictated by the understanding of the natural order that was dominant at that time in the scientific community. The three main museum functions - collection, research and communication (Mairesse 2002) - were during this period collapsed in one and only function: the collection that was exhibited to the public was identical to the one that was preserved by the museum, and also identical to the one that was studied by museum researchers.

In this context, dioramas constituted a great innovation compared to the classical museum practices. Using the logic of analogy, diorama builders stated the intention of "representing" or "imitating "nature in a realistic manner (Anderson 2015; Montpetit 1996). The first dioramas were showing groups of selected objects and specimens in a certain context, with the intention of allowing "people with no access to nature to see animals exhibited in their natural habitat" (Ash 2004, p. 86) and inducing a certain kind of learning (Bitgood 1996). Their purpose was educational from the start, without any relation with the museum collection/research functions (Montpetit 1996).

As the first generation of dioramas were mostly conceived by taxidermists and/or artists, and only rarely by scientists, they were for the majority of the time representing the personal perceptions of those individuals about nature, who also wanted to share their love of nature with others (Howie 2015). Although described as works of art, dioramas were nonetheless meant to be truthful to the original scene, "they were meant to be accurate." (Anderson 2015, p. 72). With time, these initial personal efforts were progressively institutionalized and structured around the work of multidisciplinary teams, including exhibition designers, scientists, artists, taxidermists - making the represented scenes more scientifically accurate (Marandino et al. 2015).

Nowadays, researchers interested in studying the educational role of dioramas identify many different intentions driving both their conceptions by designer teams and uses by museum educators. Sometimes, dioramas are presented in museums with the simple intentions of representing nature (Howie 2015; Morris 2015). Other times, they are meant to contextualize (Bitgood 1996; Montpetit 1996), to enable visitors to understand the dynamic interactions occurring within ecosytems (Tunnicliffe and Scheersoi 2015), to develop aesthetic awareness, to catch visitors' attention and stimulate their imagination (Bitgood 1996). Or, they are meant to induce specific attitude towards nature (Anderson 2015) such as "to promote an ethic for the preservation of species and their habitats" (Ash 2004, p. 84), through an immersive-like experience inducing specific emotions (Bitgood 1990; Montpetit 1996; Tunnicliffe and Scheersoi 2015). Finally, dioramas are meant to induce cognitive changes, not only on the form of acquisition of factual knowledge, but also in provoking a wide range of learning behaviors such as: collective sense-making (Ash 2004), as well as inquiring, questioning, naming and interpreting (Tunnicliffe 2015). From this short review of the literature, we can see that the educational intentions behind dioramas are multiple and varied. However, when we did look for research specifically interested in models and modeling, we have found only one paper that mentioned the educational role of dioramas taken as models (Loveland et al. 2015). For that reason, we wish here to deepen this reflection in order to better understand the possible contribution of dioramas in the learning about models and modeling. In this purpose, before discussing the educational role of dioramas as models, we will first define what models are and what the purpose of modeling in science is.

7.3 Models and Modeling

Many different objects, such as toy cars, computer simulations, and formulas, can be described as models. Something can be said to be a model, from the moment this something is selected and used by someone in order to achieve a certain goal (Mahr 2008). One object may accordingly be a model or not, depending on the circumstances under which it is viewed (Stachowiak 1973). Models can also be found and used in everyday life. Indeed, people meet models in many situations, whether at their dentist while deciding about treatments, when looking at the weather forecasts, when reading a book about Romans' canalization.

Despite this broad concept of the term "model", there is one thing all models have in common: they are carriers of ideas and conceptions (Mahr 2009). In the literature about models, there are numerous definitions of the concept, which vary according to the research fields and often bring scientists in arguments with each other. In his general theory about models, Stachowiak (1973) describes the three main features of a model. The first one, the function of *representation*, refers to the fact that models, by representing or copying an original of something, can replace this something. Models can be said to be "a model from something". However, a model is not meant to show the reality as it really is. Instead it bears the function of *reduction*. Depending on the person who's creating or using them, on the goal this person is pursuing, or on the period of time during which they have been created or used, models will present different features. Indeed, models are contextual: they are *pragmatic* in nature.

These contextual conditions also affect the nature and roles of models exhibited in natural history museums. As such, models play different roles in the museum depending on who is using them and for what reason. At the museum, scientists use models as experimental objects or instruments with the goal of producing new scientific knowledge. Museum designers and educators use models in order to communicate facts, ideas, questions, or scientific concepts to the visitors. Finally, museum visitors will use museum models for their personal learning needs. Thereby, each and everyone of those 'actors' brings a new perspective on museum models (Moormann and Bélanger 2015).

Furthermore, thinking with the help of models allows individuals to reflect on and to discuss about complex matters (Clement 2000). Indeed, in the cognitive sciences, models have been described to exist not only in the form of real three-dimensional objects or machines, but also in the form of mental representations of complex ideas or systems (Clement 2000). Mental models can be defined as internal representations of objects, situations or processes a person uses to explain how the world functions. Mental models enable individuals to make predictions, inferences or decisions (Gilbert and Priest 1997). In this regard, Gilbert and Priest (1997) describe three types of mental models. There are "expressed models" which are a version of a mental model externalized by an individual through action, speech or writing. There are "consensus models" which are expressed models that have been subjected to testing and evaluation of a community, such as the scientific community.

Finally, there are "teaching models" which are specifically constructed with the purpose of teaching and learning about a certain consensus model and developing students' understanding of this model (Treagust et al. 1998).

7.3.1 Models and Modelling in the Natural Sciences

In the context of the natural sciences, models play a central role in the scientific ways of thinking and working (Upmeier zu Belzen and Krüger 2010). Scientific models serve as tools for scientists to exchange scientific knowledge (Mittelstraß 2004). Also, Schwarz et al. (2009) define a scientific model "as a representation that abstracts and simplifies a system by focusing on key features to explain and predict scientific phenomena" (p. 633). A model consists of "elements, relations, operations, and rules governing interactions expressed by using external notation systems." (Lesh and Doerr 2003 p. 10). The elements are used to represent important aspects of a natural phenomenon. Therefore they are conceptual elements (Schwarz et al. 2009).

Gilbert (1995) defines scientific work as a construction of models which represent ideas about the world. Accordingly, scientists use models as means for thinking and for developing new understandings about the world. This is achieved through a "cyclical process of hypothesis generation, rational and empirical testing, and modification or rejection." (Clement 2000, p. 1050). Indeed, models are generative in nature and most scientists look at models as being highly dynamic and alterable (Schwarz et al. 2009). In the literature, the thinking that occurs from the cyclical construction, evaluation and revision of models has been referred to as "model-based reasoning" (Lehrer and Schauble 2000). Considered as one of the main scientific practices, modeling is a great source of creative changes in science (Lorenzo et al. 1999).

7.3.2 Models in Science Education

With the major reforms that have happened around the world during the last decade, science education now generally pursues the goal, not only to train future scientists, but also to develop scientific literacy of all students. Because the use of models is recognized as one of the central practices associated with the natural sciences, the learning about models has become part of the educational standards of many countries, including UK, Germany, Canada, and United States (DFEE 1999; KMK 2005; MELS 2006; NRC 2013). In fact, models play an active part in each of the three aspects of scientific literacy, as described by Hodson (1993). In science classes, students get to learn the major models that were produced by scientists. By creating and testing their own models in science laboratories, students are doing science, and they get to learn about the functions of models in scientific activity (Justi and Gilbert 2003; Orsenne 2016). Thus, without models the natural sciences cannot be learned, nor can they be taught (Harrison and Treagust 2000).

However, research about models in science education has shown that students often perceive models as replicas of reality with the purpose of showing objects. They have a limited understanding of models as being miniature depictions of real-life objects instead of instruments used within an epistemological process (Grosslight et al. 1991; Treagust et al. 2002; Meisert 2008; Trier et al. 2013). One reason could be that in the teaching and learning context, the descriptive aspect of models is predominant, whereas their heuristic function as thinking and working tools is not recognized. Consequently, the competent *use of models* as well as the ability to *think in models* - which is referred to as "model competence" - is an open-door to understanding the nature of science and scientific knowledge (Leisner 2005). In this regard, Upmeier zu Belzen and Krüger (2010) have developed a normative theoretical structure describing conceptions, skills and abilities linked to the development of model competence. This model entails two dimensions: the cognitive dimension 'knowledge about models' and the practical dimension 'modeling', comprising each several subcomponents. A detailed description of the theoretical structure, its components and subcomponents, can be found at Grünkorn et al. (2014). This theoretical model has proven to be useful in designing teaching and learning situations in class, as well as in studying the development of students' model competence. It bears the potential to help the work of museum educators who are willing to use museum models, such as dioramas, in their educational interventions.

7.3.3 Are Dioramas Scientific Models?

From the above, we are now able to answer the first question we wanted to address in this text: *To what extent can dioramas be considered scientific models?* We have seen that scientists use models to illustrate or visualize complex ideas or systems. They also use models to communicate their ideas to others, as well as to test and refine their understandings during a progressive and collective process which leads to the generation of new scientific knowledge. In museums, we have seen that dioramas are mainly created by artists, taxidermists or designer teams, which sometimes include scientists. As such, dioramas are considered to be visual representations of the scientific models that have inspired them. However, once they are presented in museums exhibition rooms, dioramas are not usually used by scientists in their work. Therefore, as dioramas are mainly designed to educate visitors and not for the purpose of pursuing scientific research, we conclude that they can be said to be educational models, but not scientific models. The specific nature and roles of dioramas as educational models will be discussed in the next section.

7.4 Museum Dioramas to Promote Model-Based Learning

In the educational sciences, Clement (2000) has described how students' mental models can be brought to progressively change and become more complex by the use of didactic models in class (Figure 7.2a). Starting from an initial mental model

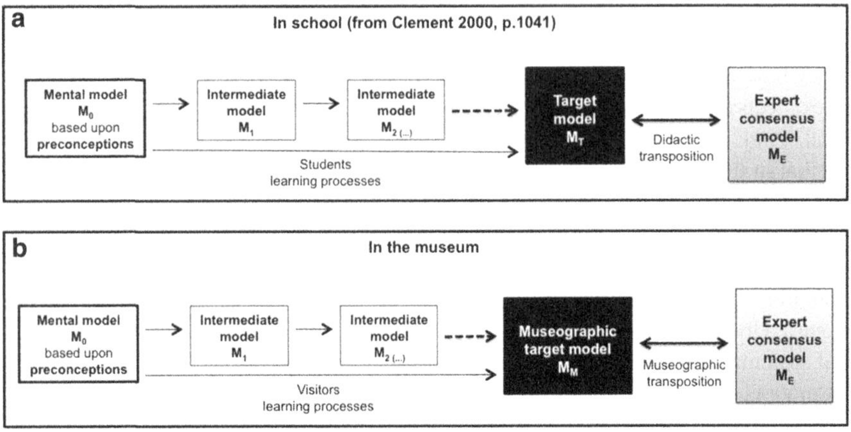

Fig. 7.2 The model-based learning frameworks in school and in the museum

(M_0) - built upon the preconceptions about a certain concept, theory, or system - it slowly evolves first to another somewhat more complex model (M_1) then to a second one (M_2) and to another (M_3)... until the mental model equals the target model (M_T) that was set by a teacher as a goal for the learning activity.

According to Clement (2000), the target model (M_T) entails the transformation of scientific knowledge into a certain kind of knowledge that is more accessible to the students, according to their age or level at school (Figure 7.2a). The target model (M_T) is related to the expert consensus model but "may not be as sophisticated as the expert consensus model currently accepted by scientists." (Clement 2000, p. 1042). It is thus a transformed and simplified visual representation of a selected scientific model (M_E) which is used in class as a teaching tool. Depending on the level of the students the target model (M_T) will be more or less similar to the scientific model (M_E) it exemplifies. This model-based learning framework can be made useful in science classes with the goal of teaching new complex knowledge (such as function, change, or interdependence) and also for teaching about the cyclic modeling process itself, in order to encourage the students to understand better the scientific way of constructing new knowledge.

Starting from this model-based learning framework developed by Clement (2000) for the learning occurring in school, we have derived a similar framework describing the learning process that occurs when museum objects and exhibits are used as models. In our recent research, the museum-derived framework (Figure 7.2b) has been useful to conceptualize the learning process that occurs in this specific environment of the museum.

The framework allows us to answer the second question we had put forth in the introduction, which was: *How can dioramas promote model-based learning?* The process starts from the act of museum designers and educators who transform selected scientists' consensus models (M_E) into museographic target models (M_M). This is achieved through the process of "museographic transposition" which has been described elsewhere (cf. Mortensen 2010; Simonneaux and Jacobi 1997). The framework shows that the encounter of the simplified representations of more complex models in the museum can induce model-based learning and the gradual development of new conceptual understandings about those models.

In the specific case of dioramas, because they are the product of the act of museographic transposition performed by museum designers and educators, they are the "external representations of the consensus models of scientists and of the mental models of their designers." (Norman 1982, cited in Loveland et al. 2015, p. 94). From there, the framework describes the learning process that occurs when visitors interact with dioramas taken as museographic models, which are set as learning goals. It shows how mental models can gradually evolve (M_1, M_2, M_3, …) when confronted to more sophisticated ones, towards the targeted museographic models.

In our recent research work, we have started exploring the learning about models that occurs during a school visit at the Museum für Naturkunde Berlin. Using concept maps and group discussions, we have been able to describe different forms of conceptual change with regard to students' understandings about models that result from a school visit to the museum (Moormann and Bélanger 2017). In our future work, we intend to go further in this exploration of model-based learning in the museum. Using the museum-derived framework, we plan to look at the gradual transformation of the students' mental models about natural ecosystems through repeated and varied encounters with dioramas which we consider, taken as the targeted museographic models.

7.5 Conclusion

In this chapter, our intention was to discuss the possible role of museum dioramas as models in promoting model-based learning. In order to do so, we went back to the history of dioramas to identify the purposes underlying their design and exhibition in natural sciences museums. We then presented a short review of the literature about models and modeling, with a focus on their uses in the natural sciences and in science education. This discussion brought us to the conclusion that: because museum dioramas are meant to promote learning, and, because they are not normally used by scientists once they have been put in the exhibition halls, dioramas can be said to be educational models, but not scientific models. They are a product of the "museographic transposition" process, which results in simplified external representations of selected scientists' consensus models combined with the personal mental models of the museum designers or educators. As such, inspired by the work of Clement (2000), we have inferred that dioramas have the potential to promote model-based learning by playing the role of target models that gradually may bring the visitors' mental models to become more and more complex, in the direction of the scientists' consensus models. The research that we intend to pursue in the near future, using the museum-derived model-based learning framework outlined in this paper, should bring a better understanding on the effect of dioramas on visitors in promoting learning with regards to biological phenomena and natural ecosystems.

References

Anderson, M. (2015). James Perry Wilson: Shifting paradigms of natural history diorama background painting. In S. D. Tunnicliffe & A. Scheersoi (Eds.), *Natural history dioramas history, construction and educational role* (pp. 67–78s). Dordrecht: Springer.

Ash, D. (2004). How families use questions at dioramas: Ideas for exhibit design. *Curator The Museum Journal, 47*(1), 84–100.

Bitgood, S. (1990). Toward an objective description of the visitor immersion experience. *Visitor Behavior, V*(2), 11–14.

Bitgood, S. (1996). Les méthodes d'évaluation de l'efficacité des dioramas: compte rendu critique. *Publics et Musées, 9*(1), 37–53. https://doi.org/10.3406/pumus.1996.1070.

Borg, J. J. (2015). Dioramas in Natural History Museum—Tools for nature conservation. In S. D. Tunnicliffe & A. Scheersoi (Eds.), *Natural history dioramas history, construction and educational role* (pp. 81–86). Dordrecht: Springer.

Clement, J. (2000). Model based learning as a key research area for science education. *International Journal of Science Education, 22*(9), 1041–1053. https://doi.org/10.1080/095006900416901.

Damaschun, F., Hoff, N., & Spieler, M. (Eds.). (2010). *Museum für Naturkunde the exhibitions*. Berlin: Museum für Naturkunde.

Department for Education and Employment – DFEE. (1999). *The National Curriculum handbook for primary teachers in England*. London: Qualifications and Curriculum Authority.

Fortin-Debart, C. (2003). Le musée de sciences naturelles, un partenaire de l'école pour une éducation relative à l'environnement: du message scientifique au débat de société. *VertigO La revue des sciences de l'environnement, 4*(2), 1–10.

Gilbert, J. K. (1995). Learning in museums: Objects, models and text. *Journal of Education in Museums, 16*, 16–18.

Gilbert, J. K., & Priest, M. (1997). Models and discourse: A primary school science class visit to a museum. *Science Education, 81*, (6), 749–(6), 762.

Girault, Y. (2003). Le musée de science: d'un parti pris épistémologique à la prise en compte des publics. In Y. Girault (Ed.), *L'Accueil des publics scolaires dans les Museums - Aquariums, Jardins botaniques, Parcs Zoologiques* (p. 296). Paris: L'Harmattan.

Girault, Y., & Sauvé, L. (2008). L'éducation scientifique, l'éducation à l'environnement et l'éducation pour le développement durable - Croisements, enjeux et mouvances. *Aster, 46*, 7–30.

Grosslight, L., Unger, C., & Jay, E. (1991). Understanding models and their use in science: Conceptions of middle and high school students and experts. *Journal of Research in Science Teaching, 28*(9), 799–822.

Grünkorn, J., Upmeier zu Belzen, A., & Krüger, D. (2014). Assessing students' understandings of biological models and their use in science to evaluate a theoretical framework. *International Journal of Science Education, 36*(10), 1651–1684. https://doi.org/10.1080/09500693.2013.873155.

Harrison, A. G., & Treagust, D. F. (2000). A typology of school science models. *International Journal of Science Education, 22*(9), 1011–1026.

Hodson, D. (1993). Re-thinking old ways: Towards a more critical approach to practical work in school science. *Studies in Science Education, 22*(1), 85–142.

Howie, G. (2015). Chapter 5 Dioramas as constructs of reality: Art, photography, and the discursive space. In S. D. Tunnicliffe & A. Scheersoi (Eds.), *Natural history dioramas - history, construction and educational role* (pp. 39–65). Dordrecht: Springer.

Justi, R. S., & Gilbert, J. K. (2003). Teachers' views on the nature of models. *International Journal of Science Education, 25*(11), 1369–1386. https://doi.org/10.1080/0950069032000070324.

KMK. (2005). *Bildungsstandards im Fach Biologie für den Mittleren Schulabschluss*. München Neuwied: Wolters Kluwer.

Köstering, S. (2003). *Natur zum Anschauen* (pp. 1871–1914). Köln: Das Naturkundemuseum des Deutschen Kaiserreiches Böhlau.

Köstering, S. (2015). Heimatgefühle in 3-D. In A. Hermannstädter, I. Heumann, & K. Pannhorst (Eds.), *Wissensdinge. Geschichten aus dem Naturkundemuseum Berlin* (pp. 160–163). Berlin: Nicolai.

Lehrer, R., & Schauble, L. (2000). Developing model-based reasoning in mathematics and science. *Journal of Applied Developmental Psychology, 21*(1), 39–48.

Leisner, A. (2005). Modellkompetenz im Physikunterricht. In H. Giest (Ed.), *Lern- und Lehrforschung: LLF-Berichte 20* (pp. 35–50). Potsdam: Universitätsverlag Potsdam.

Lesch, R., & Doerr, H. M. (2003). Foundations of models and modeling perspective on mathematics teaching, learning, and problem solving. In R. Lesch & H. M. Doerr (Eds.), *Beyond constructivism: Models and modeling perspectives on mathematics problem solving, learning, and teaching* (pp. 3–33). Mahwah: Earlbaum.

Lorenzo, M., Nersession, N. J., & Thagard, P. (1999). *Model-based reasoning in scientific discovery*. Dordrecht: Kluwer Academics.

Loveland, M., Buckley, B. C., & Quellmalz, E. S. (2015). Chapter 8 Using technology to deepens and extend visitors' interactions with dioramas. In S. D. Tunnicliffe & A. Scheersoi (Eds.), *Natural history dioramas history, construction and educational role* (pp. 87–104). Dordrecht: Springer.

Mahr, B. (2008). Ein Modell des Modellseins. Ein Beitrag zur Aufklärung des Modellbegriffs. In U. Dirks (Ed.), *Modelle* (pp. 187–218). Frankfurt am Main: Lang.

Mahr, B. (2009). Information science and the logic of models. *Software& Systems Modeling, 8*, 365–383.

Mairesse, F. (2002). *Le Musée temple spectaculaire*. Lyon: Presses universitaires de Lyon.

Marandino, M., Achiam, M., & Dias De Oliveira, A. (2015). The diorama as a means for biodiversity education. In S. D. Tunnicliffe & A. Scheersoi (Eds.), *Natural history dioramas history, construction and educational role*. Dordrecht: Springer.

Meistert, A. (2008). Vom Modellwissen zum Modellverständnis – Elemente einer umfassenden Modellkompetenz und deren Fundierung durch lernerseitige Kriterien zur Klassifikation von Modellen. *Zeitschrift für die Didaktik der Naturwissenschaften, 14*, 243–261.

Ministère de l'Éducation, du Loisir et du Sport – MELS. (2006). *Le Programme de formation de l'école québécoise*. Québec: Ministère de l'Éducation, du Loisir et du Sport, Gouvernement du Québec.

Mittelstraß, J. (Ed.). (2004). *Enzyklopädie Philosophie und Wissenschaftstheorie Bd. 2* (pp. 911–914). Stuttgart: Verlag J. B. Metzler.

Montpetit, R. (1996). Une logique d'exposition populaire: les images de la muséographie analogique. *Publics et Musées, 9*(1), 55–103. https://doi.org/10.3406/pumus.1996.1071.

Moormann, A., & Bélanger, C. (2015). Effect of a school visit to a Natural History Museum on pupils' conceptions about models in biology. Paper presented at the 11th International Conference of the European Science Education Research Association (ESERA), Helsinki, Finnland, 31 August 4 September 2015.

Moormann, A., & Bélanger, C. (2017). Entwicklung von Schülervorstellungen zu Modellen im Kontext eines Schulbesuchs im Museum für Naturkunde. In J. Stiller & C. Laschke (Eds.), *Berlin-Brandenburger Beiträge zur Bildungsforschung 2017, Herausforderungen, Befunde und Perspektiven interdisziplinärer Bildungsforschung* (pp. 205–229). Frankfurt am Main: Peter Lang.

Morris, P. (2015). Chapter 4 A window on the world—Wildlife dioramas. In S. D. Tunnicliffe & A. Scheersoi (Eds.), *Natural history dioramas history, construction and educational role* (pp. 33–37). Dordrecht: Springer.

Mortensen, M. F. (2010). Museographic transposition: The development of a museum exhibit on animal adaptations to darkness. *Éducation & Didactique, 4*(1), 115–138.

National Research Council – NRC. (2013). *Next generation science standards*. Washington, DC: National Academies Press.

Orsenne, J. (2016). *Aktivierung von Schülervorstellungen zu Modellen durch praktische Tätigkeiten der Modellbildung*. Dissertation, Humboldt Universität zu Berlin.

Scheersoi, A. (2014). Warum Dioramen aus didaktischer Perspektive so wertvoll sind. In Schulze, A., Köhler, J., & Gruber, G. (Eds.), Naturkundliche Dioramen. *Kaupia - Darmstädter Beiträge zur Naturgeschichte 19*: 121–126.
Schwarz, C. V., Reiser, B. J., Davis, E. A., Kenyon, L., Achér, A., Fortus, D., Shwartz, Y., Hug, B., & Krajcik, J. (2009). Developing a learning progression for scientific modeling: Making scientific modelling accessible and meaningful for learners. *Journal of Research in Science Teaching, 46*(6), 632–654.
Simonneaux, L., & Jacobi, D. (1997). Language constraints in producing prefiguration posters for a scientific exhibition. *Public Understanding of Science, 6*, 383–408.
Stachowiak, H. (1973). *Allgemeine Modelltheorie*. New York: Springer.
Treagust, D. F., Harrison, A. G., & Venville, G. J. (1998). Teaching science effectively with analogies: An approach for preservice and inservice teacher education. *Journal of Science Teacher Education, 9*(2), 85–101.
Treagust, D., Chittleborough, G., & Mamiala, T. L. (2002). Students'understanding of the role of scientific models in learning science. *International Journal of Science Education, 24*(4), 357–368.
Trier, U., Upmeier zu Belzen, A., & Krüger, D. (2013). Students' versus scientists conceptions of models and modelling. In M. Ekborg, D. Krüger, D. J. Boerwinkel, M. Ergazaki, M. J. Gil Quilez, & G. Molinatti (Eds.), *Proceedings of the 9th conference of European researchers of didactics of biology* (Vol. 2012, pp. 103–115). Berlin: ERIDOB.
Tunnicliffe, S. D. (2015). Chapter 13 Naming and narratives at natural history dioramas. In S. D. Tunnicliffe & A. Scheersoi (Eds.), *Natural history dioramas history, construction and educational role* (pp. 161–186). Dordrecht: Springer.
Tunnicliffe, S. D., & Scheersoi, A. (2015). Chapter 11 dioramas as important tools in biological education. In S. D. Tunnicliffe & A. Scheersoi (Eds.), *Natural history dioramas - history, construction and educational role* (pp. 133–143). Dordrecht: Springer Science+Business Media.
Upmeier zu Belzen, A., & Krüger, D. (2010). Modellkompetenz im Biologieunterricht. *Zeitschrift für die Didaktik der Naturwissenschaften, 16*, 41–57.
Van Praët, M. (1989). Contradictions des musées d'histoire naturelle et évolution de leurs expositions. In B. Schiele (Ed.), *Faire voir, Faire savoir. La muséologie scientifique au présent* (pp. 25–34). Musée de la civilisation: Québec.

Alexandra Moormann is a certified biologist (physiological ecology) and a trained teacher in biology and physics. She has done her Ph.D. in Biology Education at Humboldt-Universität zu Berlin in Germany about the development of students' attitudes towards science subjects with the focus on transitions in a longitudinal design. For more than 15 years, she has worked as a freelance museum educator at a botanical garden and a natural history museum. She now works at the Museum für Naturkunde Berlin as a researcher in museum education. Her research interests are "models in museums "and "Education for Sustainable Development (ESD) in natural history museums with participatory tools".

Charlène Bélanger After completing a Ph.D. in Educational Sciences at Université du Québec à Montréal, she currently holds the position of Projection Manager, Education at the McCord Stewart Museum in Montreal, Canada. Her research interests focus on collaborative learning. Using ethnographic and qualitative methods, she has studied the teaching and learning of scientific practices, such as argumentation or model-based reasoning. Her research, carried out both in schools and museums, is situated at the junction between museum, environmental, and citizenship education.

Chapter 8
Educational Mechanisms of Dioramas

Michael May and Marianne Achiam

8.1 Meeting the Objects

Designers and curators tend to take their own perspective on the museum experience and conceptualize museum visits as communication: an indirect communication of design intentions to visitors mediated by exhibits, labels, explanatory signs, and narratives. From the perspective of the museum visitor, however, the basic level experience can best be described as a meeting with artefacts or natural objects on display. Museum artefacts and objects may involve interaction and dialogue, but usually they are not directly communicative unless they take the form of guided tours, multimedia installations or live animals. Rather than a form of communication we will frame the basic level experience of museum visitors as a form of inquiry.

Meeting artefacts and object in the museum context is bound to raise implicit questions of identification corresponding to explicit questions such as "what is this?" The interest and motivation of museum visitors will also orient their attention, but embodied perception and basic level categories are generally at the core of human experience (Evans and Green 2006). At a basic level of observation visitors may have enough prior knowledge about an exhibition and its subject matter to identify objects immediately.

The inquiry aspect of a museum visit will arise when visitors are confronted with exhibited objects they cannot immediately identify or situations they do not understand. This can set in motion a sequence of actions such as looking for information

M. May (✉)
Southern University of Denmark, Kolding, Denmark
e-mail: mma@sdu.dk

M. Achiam
University of Copenhagen, Copenhagen, Denmark
e-mail: achiam@ind.ku.dk

A. Scheersoi, S. D. Tunnicliffe (eds.), *Natural History Dioramas – Traditional Exhibits for Current Educational Themes*,
https://doi.org/10.1007/978-3-030-00175-9_8

Fig. 8.1 Oryx gazelle diorama from the American Museums of Natural History. Image courtesy of Dano, licensed through Creative Commons. Link to material: https://www.flickr.com/photos/mukluk/440494699. No alterations to the image have been made

on labels or explanatory signs next to a diorama. Our focus here is however on the experience of and learning from dioramas in natural history museums (although we note that dioramas also occur in anthropological and technical museums, and that similar educational mechanisms may be at play in these settings). As a naturalistic or pseudo-naturalistic display of objects (e.g. humans, animals, plants, machines) in their natural or cultural environment, dioramas and their interpretation can appear straightforward, but complexity hides behind the naturalistic surface.

Consider the African Oryx gazelle (Gemsbok in Afrikaans) diorama in Fig. 8.1 from the American Museum of Natural History in New York. Visitors might find this static display of gazelles rather uneventful. We immediately recognize five similar gazelles assembled on an African savannah, and we will probably focus at first on their impressive horns. If we dwell in front of the diorama, we might, however, be more intrigued or even disturbed by the scene. The gazelles are not just on display as in a static image: we discover that they are actually looking back at us! We know that we are just looking at artefacts, but for a moment we are propelled into the scene on the savannah, as if we were really there as participants in an interspecies meeting. We have created an *imagined event* from the static display of artefacts, but how is this possible? Before discussing this we will introduce some concepts from perceptual psychology and cognitive linguistics in order to analyse how we recognize and interpret visual scenes.

8.2 The Gestalt Laws of Perception

Even though the concept of Gestalt is commonly used to describe perceptual patterns, the origin and nature of the Gestalt laws are not generally appreciated. Gestalt psychology gradually developed as a reaction to the focus of early experimental psychology on elementary sensations in the beginning of the twentieth century. Perceptual psychology had been pioneered by physicists, among others Herman von Helmholtz (1821–1894) and Gustav Theodor Fechner (1801–1887), but in the specific form of a *psychophysics* of sensations. Inspired by phenomenology, Carl Stumpf (1848–1936) founded an institute for experimental psychology in Berlin around 1900, and the movement known as Gestalt psychology was formed by his students Max Wertheimer (1880–1943), Kurt Koffka (1886–1941), Wolfgang Köhler (1887–1967) and Kurt Lewin (1890–1947). In reaction to the elementary sensations studied by classical experimental psychology the Gestalt psychologists claimed that the perceptual experience of humans and other animals was given in the form of structured wholes, and that these 'Gestalten' had their own inner laws of organization and dynamics (Ash 1998, 220). The first Gestalt laws were formulated by Max Wertheimer in the early 1920s, and included the basic principles of *proximity, similarity, closure* and *good continuation*, and Köhler made German Gestalt theory known internationally through its opposition to introspection as well as to behaviourism (Köhler 1947).

The basic Gestalt principles are still valid today, and widely utilised in information visualization and interface design (Palmer 1992; Ware 2000). According to the *proximity* principle we will experience objects that are spatially close as meaningfully grouped together, and according to the *similarity* principle we will experience objects that are visually similar in shape, size or colour as meaningfully grouped together. *Closure* and *good continuation* are related phenomena where we tend to supplement perceived curves and shapes in order to identify them as simple geometric figures. We will for instance recognize a broken ring as a circle and a broken line as a line, and we will similarly tend to look for simple geometric shapes in the contours of objects, even if we have to supplement with our own imaginary contours. These principles of closure and continuation are important for naturalistic perception. Notice how we will naturally perceive the partially occluded gazelle in the background of the diorama (Fig. 8.1) as a whole gazelle, and not as a gazelle head severed from the rear part of the animal! Furthermore, we clearly recognize the gazelles as a group closely related by spatial proximity, and – as stressed by the Gestalt psychologists – we experience all these phenomena as *given* in perception and not as assembled or constructed from elementary sensations.

This leads us to another aspect of the perceived scene, namely its organization into foregrounded *figures* and the *background* of the visual scene. Ambiguities of the figure-ground organization were discovered by the Danish psychologist Edgar Rubin (1886–1951) and further discussed by Köhler, but the fundamental figure-ground phenomena is the separation itself, i.e. the foregrounding of figures in perception. In order to work out how the Gestalt principles of perception work as part of a more comprehensive interpretation of visual scenes, we have to consider how they interact with embodied cognition and language. This is explicitly addressed in cognitive linguistics where the grammatical structures of natural languages can be seen to add more structure to the organizing principles of perception (Evans and Green 2006).

8.3 Schematic Meaning in Grammar and Cognition

One of the fundamental assumptions in cognitive linguistics is the idea that grammar is not a formal syntactic system but carries meaning beyond the individual words in phrases. This is stressed in different traditions of cognitive linguistics such as cognitive semantics (Talmy 2000) and cognitive grammar (Langacker 2013). The grammatical meaning in conveyed by word classes (nouns, verbs, prepositions, adjectives etc.) and by grammatical forms such as metaphorical constructions and other types of schematic and idiomatic expressions. Nouns represent objects, verbs events, prepositions relations, and adjectives properties, and as such they carry schematic meanings beyond the individual words.

Another fundamental assumption in cognitive linguistics is the idea that spatial representation and spatial reasoning plays an important role in binding together language, cognition, and action by sharing schematic meaning. An example is the figure-ground effect in perception that is elaborated in natural language by linguistic operations that will correspondingly *profile* objects and events in the foreground of a described scene and push other aspects of a situation in the background. Another example is that we will always construe a described situation from a particular *perspective*, just as we will always experience a scene perceptually from a particular physical perspective.

It is because perception and language share schematic meanings that it is possible to 'translate' between them. Otherwise it would be a mystery how the visual structure of a situation could somehow be related to structures in a language. We do, however, find it quite easy to describe the scene of a situation we are experiencing or conversely, to visualize in our imagination the scene of a situation being described to us.

Consider the tiger diorama in Fig. 8.2 from the Natural History Museum in Helsinki. Visitors can move past the diorama and change their physical perspective from e.g. behind the tiger to a perspective in front of the deer being attacked. This will correspond to a change in relevant descriptions from the perspective of the tiger as grammatical subject (e.g. "The tiger attacks the dear") to a passive construction where the deer is the grammatical subject (e.g. "The deer is attacked by the tiger"). The most striking feature of the diorama is clearly that the static scene is experienced as a dramatic *event*. The Gestalt psychologist was in fact preoccupied by the dynamic nature of apparently static visual scenes as well as the *apparent motion* constructed as a Gestalt from a sequence of static images (Ash 1998). The tiger diorama seems to take us a step further though: even though we only see a single

Fig. 8.2 Tiger diorama at the Natural History Museum in Helsinki, Finland. Image courtesy of Daderot, licensed through Creative Commons. Link to material: https://commons.wikimedia.org/wiki/File:Tiger_diorama_-_Finnish_Museum_of_Natural_History_-_DSC04665.JPG. No alterations to the image have been made

'frozen' image of the apparent movement, we are compelled to *imagine the whole event* of the tiger's leaping jump and the attempted escape of the deer. We do not need a series of 'snapshot' scenes in order to complete the movement (cf. the Gestalt effect of good continuation). We can actually 'see' the movement event in the static scene.

On the level of language this event can be elaborated in different ways according to the *perspective* we take in construing the situation, i.e. with the tiger or the deer as grammatical subjects and actors in the imagined event ("The tiger attacks the deer", "The deer is attacked by the tiger"). We can however also *profile* the situation in different ways when describing it, corresponding to the attention we can pay to different aspects of the situation. We can for instance say that "the deer is fleeing" profiling only the deer, or "the tiger is jumping" profiling only the tiger. We can even profile the background and imagine a past event ("It has been snowing"). This distribution of attention is another fundamental aspect of how meaning is constructed in language and perception (Talmy 2000).

The different possibilities of meaning construction that arise from Gestalt perception of visual scenes and their basic schematization, as well as the different choices of perspective and profiling, will simultaneously provide possibilities for dialogue and learning. The imagined event of the tiger diorama can trigger many questions concerning aspects of the situation: How can the tiger manage to jump that high? Will the deer have any chance of escaping? Many types of inquiry about the scene can depart from the diorama: about the event itself, about the species involved, about the landscape and its location, and even about more conceptual and theoretical issues derived from the scene such as predator-prey relations, the habitats of tigers, and their status as an endangered species.

8.4 Levels of Meaning

The construction of meaning made possible by the diorama is however not limited to the *lexical level* of object recognition and the *phrastic level* of scenic descriptions that we have discussed so far. Families and groups of children observing the tiger diorama will most likely engage in storytelling and dialogue. This is meaning construction at a *narrative level* of meaning (cf. Achiam et al. 2014). In this way the apparent motion of the scene will not only be the basis of an imagined event, but also the point of departure for a narration of its probable past and its possible future as well as the imagined intentions of the involved figures.

The kind of involvement we experience with the scenes and imagined events of dioramas bears witness to our power of imagination and the organizing power of language and perception. Meaning construction can be understood as proceeding on a number of levels from the perceived objects and their basic description (lexical level), to the construal of situations and events (phrastic level), and to their elaboration into stories about the past and the future as well as the intentions of displayed actors. On an even higher level of construction we have to acknowledge that the diorama as a whole, as well as the actors within it, might take on a communicative role. The diorama can have a rhetorical or "ideological" function by addressing the visitor as a spectator and as a witness to a particular scene.

Looking at the scene of the tiger diorama we are in sense caught as passive bystanders to the (imagined) dramatic event, and the diorama is thereby addressing us and imposing additional meaning associated with survival or ideological conceptions (e.g. "nature is cruel"). This we will call the *discursive* level of meaning. In semiotics this form of implicit communication that addresses us as participants is often called enunciation.

The discursive level is also operative in the Oryx gazelle diorama of Fig. 8.1, and this is what can appear as disturbing. In the moment when we realize that the group of gazelles are looking back at us, we are (so to speak) addressed by them and their gaze. This is a moment of suspension of disbelief, where we meet the other species (in our imagination). Here, the diorama is a powerful mechanism for *including the visitor in the scene*, and thus much more than a mere image or a collection of artefacts.

8.5 Learning from Dioramas

The learning potential of the diorama arises from the imaginative richness (cf. Chap. 9 by E. Mifsud) of the content it creates by placing artefacts and natural objects within a naturalistic scene (or relevant cultural context in the case of technical museums). There are, however, also several problems of learning associated with the diorama. One potential problem is the very naturalistic articulation of the diorama that creates an *apparent realism* that could entail false inferences. One typical example from museums of natural history is the educational attempt to present many species that live together in a particular habitat. In order to present several species within the frame of the diorama they are often clustered very closely in the scene, and sometimes a (false) reason for this is introduced in the scene itself, for example by distributing the animals around a water hole. In reality different species would disperse over a larger area. An example is seen in the African Plains diorama from the American Museum of Natural history (Fig. 8.3).

Some dioramas do however break off from this apparent realism by being visually more schematic. An example is the Bird cliff diorama displaying typical birds of the Faroe Islands from the Museum of Natural History in Copenhagen (Fig. 8.4).

Fig. 8.3 Close clustering of species in the *Plains diorama* in the Akeley Hall of African Mammals, American Museum of Natural History. Image courtesy of NComparato, licensed through Creative Commons. Link to material: https://commons.wikimedia.org/wiki/File:Plains_Diorama.JPG. No alterations to the image have been made

Fig. 8.4 Highly schematic *Bird cliff* diorama from the Museum of Natural History, Copenhagen. Image courtesy of Michael May

This diorama has a black and white image of the tall cliffs characteristic of the Faroe Islands in the background, but the exhibited birds in the foreground are placed on highly schematic (and purple!) platforms as a kind of visual metaphor for cliffs. Significantly, these metaphorical cliffs do not look like the image of cliffs in the background.

The absence of apparent realism in the *Bird cliff* diorama is also conveyed by the location of text within the diorama labelling and describing the different species. This disruption of apparent realism could be helpful in avoiding the potential mistakes in reasoning about the living space of different species within a habitat, but on the other hand the fascination of the naturalistic diorama is lost in this schematic display.

The learning potential of dioramas should also be understood in the larger context of the explanatory signs outside the frame of the diorama, the possible interactive mechanisms supported (like user-controlled selective spotlights), the potential of supplementary multimedia like explanatory speech, smartphone apps, ambient sound or additional video next to the diorama. Furthermore, the meaning construction of visitors will often take place within dialogues with others (families, school classes) rather than privately (in thought), and accordingly the meaning construction and inquiry will be externalised and embedded in dialogues as fragments of speech (cf. the analysis examples in Achiam et al. 2014).

8.6 Conclusion

Naturalistic habitat dioramas have been described as "windows on nature" (Quinn, cit. Kamcke and Hutterer 2015). These types of naturalistic dioramas displaying animal and plant specimens in their natural habitat were developed in association with a gradual epistemological change in the natural sciences towards an awareness of systems and systemic relations in nature, and habitat dioramas were intended to convey an appreciation of the relationships between the flora and fauna of an environment (Rader and Cain 2014; Marandino et al. 2015). It is striking that this gradual shift in the public communication of natural history in the museum context towards scientific naturalism and an awareness of ecological systems occurred in parallel with the rise of Gestalt psychology and the critique of classical experimental psychology and its focus on psychophysics of sensations and isolated stimulus-response mechanisms.

Habitat dioramas are accordingly much more than windows on nature in the basic sense of aesthetic and naturalistic displays. They provoke inquiry and reflection through the meaning construction they support. The traditional perspective on dioramas as communicating 'messages' is, however, too simplistic, as we have indicated here. It can be argued, for example, that the habitat diorama often conveys an indirect message corresponding to the intention to increase awareness about nature conservation (Kamcke and Hutterer 2015), but the reference to indirect or unspoken communication does not explain the educational mechanism of the diorama, i.e. how this interpretation of the diorama can actually take place. Our proposal has been that we can understand the educational mechanism of the diorama by referring to the principles of visual object and scene recognition as originally described by the Gestalts psychologists, and furthermore by embedding this understanding in a more comprehensive conceptualization of meaning construction as developed within cognitive linguistics (Evans and Green 2006) – specifically in cognitive semantics (Talmy 2000) and cognitive grammar (Langacker 2013).

Meaning construction from observing dioramas will take place on several levels of meaning starting with the perceptual object and scene recognition and the corresponding lexical identification of objects. Museum visitors will then be able to conceptualize the scene of a diorama on a phrastic level corresponding to simple statements in natural language about the scene in accordance with their prior knowledge about the exhibited objects (cf. Chap. 9 by E. Mifsud). This sentence-like meaning can take the form of inner thoughts or fragments of speech as part of a dialogue between visitors sharing an experience. A further level of meaning is the narrative level, where stories are constructed from the scene of the diorama. These stories are sometimes supported by added external sources such as video and explanatory diagrams, but stories can be invented by visitors even without such external support (Achiam et al. 2014). Finally, a discursive level of enunciation is required to understand how the diorama can actually address us without any direct speech, and this is where the rhetorical and ideological messages of the habitat diorama should be positioned. A detailed understanding of these levels of meaning and how they are connected would of course require a more comprehensive analysis than we have attempted here.

References

Achiam, M., May, M., & Marandino, M. (2014). Affordances and distributed cognition in museum exhibitions. *Museum Management and Curatorship, 29*(5), 461–481.

Ash, M. G. (1998). *Gestalt psychology in German culture 1890–1967. Holism and the quest for objectivity*. Cambridge: Cambridge University Press.

Evans, V., & Green, M. (2006). *Cognitive linguistics – An introduction*. Edinburgh: Edinburgh University Press.

Kamcke, C., & Hutterer, R. (2015). History of dioramas. In S. D. Tunnicliffe & A. Scheersoi (Eds.), *Natural history dioramas. History, construction and educational role*. Dordrecht: Springer.

Köhler, W. (1947). *Gestalt psychology. An introduction to new concepts in modern psychology*. New York: Liveright Publishing Corporation.

Langacker, R. W. (2013). *Essentials of cognitive grammar*. Oxford: Oxford University Press.

Marandino, M., Achiam, M., & Oliveira, A. D. (2015). The diorama as a means for biodiversity education. In S. D. Tunnicliffe & A. Scheersoi (Eds.), *Natural history dioramas. History, construction and educational role* (pp. 251–266). Dordrecht: Springer.

Palmer, S. E. (1992). Modern theories of gestalt perception. In G. W. Humphreys (Ed.), *Understanding vision. An interdisciplinary perspective*. Cambridge, MA/Oxford: Blackwell Publishers.

Rader, K. A., & Cain, V. E. M. (2014). *Life on display. Revolutionizing U.S. Museums of science and natural history in the twentieth century*. Chicago: University of Chicago Press.

Talmy, L. (2000). *Towards a cognitive semantics*. Cambridge, MA: MIT Press.

Ware, C. (2000). *Information visualization. Perception for design*. San Francisco: Morgan Kaufmann Publishers.

Michael May is an Associate Professor at the Department of Science Education, University of Copenhagen. He has a Ph.D. in Psychology and B.Sc. in Cultural Sociology. His research interests include science education, e-learning, philosophical and didactic aspects of modeling in science and semiotic aspects of formal and informal learning.

Marianne Achiam (née Mortensen) is an Associate Professor at the Department of Science Education, University of Copenhagen. She has an M.Sc. in biology and a Ph.D. in science education. Her research interests include science and particularly biology dissemination in out-of-school contexts such as museums, science centres, zoos, aquaria and botanical gardens. She is the coordinator of the Departmental research group on science education in informal settings.

Chapter 9
An Interpretation Model for Dioramas

Edward Mifsud

9.1 Introduction

Wildlife collections are not only invaluable for research purposes, but they play a central part in the education of the visiting public. Although little can rival direct interaction with authentic specimens, natural history dioramas can offer visitors opportunities to see specimens from wildlife that would otherwise be difficult to encounter. The dioramas present an ecological setting, local or foreign, through which children may appreciate the natural richness of various habitats. Natural history dioramas bring visitors closer to nature, which most may only experience in books, television, internet and other multimedia portals.

Today's largely urbanised generation has limited knowledge and interest in wildlife (Huxham et al. 2006). Natural history dioramas are particularly valuable to permit the urban community to see and possibly understand the diverse habitats with the various organisms that live within (Tunnicliffe 2005). Natural history dioramas can serve as a unique and powerful science education resource. They possess great potential for learning in Biology, particularly in aspects of biodiversity, ecological relationships and ecosystem ecology. Natural history dioramas can provide snapshots in time of past habitats and also offer the prospect to visitors to observe different habitats, categorise organisms and raise personal questions. 'At their best they are one of the most powerful techniques for emotional access and effective learning' (Insley 2007).

Natural history dioramas remain an underutilised educational resource and have been dismissed as old fashioned and irrelevant by non-educator management officials enticed by effective technological innovations. Discussions that occur at dioramas embody basic science processing skills: observing, communicating, classifying,

E. Mifsud (✉)
University of Malta, Msida, Malta
e-mail: edward.mifsud@um.edu.mt

A. Scheersoi, S. D. Tunnicliffe (eds.), *Natural History Dioramas – Traditional Exhibits for Current Educational Themes*,
https://doi.org/10.1007/978-3-030-00175-9_9

inferring, and hypothesizing. Skilfully constructed natural history dioramas can still provide a significant opportunity for fundamental acquisition of science knowledge (Stern 2009).

9.2 Children Perceiving Biota

Various researchers have recently documented the educational potential and role in biological learning of dioramas (Ash 2004; Insley 2007; Peart and Kool 1988; Pipueras et al. 2008; Reiss and Tunnicliffe 2007; Scheersoi 2009; Tunnicliffe 2002, 2005). At dioramas, learning occurs through imagery in the iconic mode, which is a 'more concrete way of learning' (Hooper-Greenhill 1994). These displays potentially provide precious opportunities for education in museums (Paddon 2009).

When people look at biological exhibits in a science museum, a botanic garden or a zoo they construct meaning from what they observe whatever it may be (Bruner et al. 1956). When people encounter animals and plants in places such as parks, zoos, farms, nature reserves and natural history museums, they interpret what they see and try to make sense of it. Tunnicliffe (2002) found that primary school and family visitors to zoos and museums have a need to identify specimens using non-taxonomic basic terms. They usually hold a basic concept of the animal, which leads them to make remarks on size and anatomy. They also comment on behavioural aspects such as; position in the exhibit, locomotion, feeding and other activities that attract observer attention such as parental care (Tunnicliffe 2002). Reiss and Tunnicliffe (2007) investigated how children aged 5–14 years recognize, identify and group animals. The majority of children gave anatomical rather than behavioural or habitat reasons for naming and explaining animals. Huxham et al. (2006) found in the UK greater knowledge about mammals than about birds and arthropods. Interestingly, apart from a few local species, most Maltese children mentioned a variety of non-endemic species such as tiger, lion, crocodile, giraffe, shark and leopard (Tunnicliffe et al. 2008).

American children were more interested in endemic animals compared to national or international animals (Patrick and Tunnicliffe 2011; Trowbridge and Mintzes 1985, 1988). In a study conducted in Brazilian schools, found that 6–8 year olds mainly mentioned earthworm, bird, butterfly, toad and pigeon as animals they see around them. As pets they mentioned dog, cat and fish, and hen, duck, goat and horse as farm animals. When probed on knowledge about insects, Brazilian 4–6 year olds could spontaneously recall very few names among which were butterfly, bee, beetle, ant and surprisingly caterpillar (Bartoszeck et al. 2009). The results of these studies show that, due to cultural variances, children from different countries tend to vary in the way they interpret the concept 'animal'.

Gatt et al. (2007) investigated 4–5 year old children's knowledge and exposure to plants, and the conceptual framework used to classify a specimen as a plant. When asked to mention plants, a third of the pupils in the study did not even give a single name and very few were able to mention more than three examples. Children most frequently gave the super-ordinate categories 'flower', 'tree' or 'plant' and a few mentioned 'rose' and 'sunflower'. Most commonly mentioned trees were the orange and the apple followed by lemon. For many pupils one characteristic sufficed to classify the specimen as a plant. Specimens that fitted in the mental model were classified as plants; the example of lettuce fitted because it was green in colour, but cactus did not since it has spines in place of leaves.

At natural history dioramas children stop, look and interpret what they see, their attention captured by particular features. Such situational interest is central to learning, particularly in non-formal learning environments where visitors may be regarded as free learners (Scheersoi 2009). Dioramas stimulate situational interest if they evoke emotional responses and provide different anchor points (cf. Chap. 7 by M. May and M. Achiam). This enables visitors with varying individual backgrounds to relate previous experiences to artefacts observed. Person-object-engagements with diorama may produce feelings of enjoyment, involvement and stimulation that are typical emotional aspects of interest-based activity. Situational interest arises from: recognition of familiar, young or big animals and the unexpected (Scheersoi 2009).

As texts without readers are empty, so museum objects are bare receptacles without the agency of museum visitors. Meaning is not "put into" a text or object to be "taken away" by someone who "finds" it there, but comes into being through intersubjective participatory experiences. All participants bring certain dispositions to the encounter and no one of them has a greater claim than others to possession of the "true" meaning of the object (Hein 2000). Unlike texts, understanding objects is more complex since the categories of meaning are more ambiguous with objects than with texts and meaning is not articulated in words (Hooper-Greenhill 2000). Understanding is a process by which people match what they see and hear with pre-stored groupings of actions that they have experienced. Each person has a unique mental map of knowledge depending on prior cultural and biographical experiences. This means that each person will process new information in ways that are specific to him or her as individuals.

All objects are artefacts, a fusion between encounter and interest, irrespective of being natural or man made. On the other hand, to Hooper-Greenhill (2000) only man made things are artefacts, while the term 'specimen' is an object that belongs to the natural world (p. 106). Therefore in the classical sense a diorama, as a museum object, is an artefact, but which contains animal and plant specimens in an ecological relationship. Tunnicliffe (2013) asks whether diorama animals and plants should be considered as museum cultural objects or should they be considered as a sub-genre of objects. A new modern paradigm holds that an object's authenticity is less essential to learning when compared with its potential to support visitor participation (Eberbach and Crowley 2005). Curators of Smithsonian Institution, Spencer Crews and James Sims, declared that authenticity is located not in objects, but in the historical concepts they represent (Hein 2000).

9.3 Perception and Mental Models

Human perception occurs in two situations; (a) Natural: surfaces and textures, solid objects, rich patterns of multisensory movement and change, and (b) Human Culture: language symbols and 2D patterns as representations of 3D objects. The way in which perception engages with artefacts of our culture may differ importantly from the way in which it deals with the natural world. This becomes relevant to this research since it deals with museum artefacts, conceptual habitat dioramas, which are representations of typical local habitats (Gordon 2004).

A representation is a likeness or simulation of some ideas, concept, or object. In learning we often use an external representation, found in the environment, to build an internal representation, held in the viewer's mind. However, unlike external representations, there is no tangible evidence and we cannot physically manipulate mental representations. Very often, we must convert our mental representations into external presentations. When it is called for, we retrieve our internal representations and attempt to reproduce them in some external form (Rapp and Kurby 2008). Visualisation is of vital significance in science and science education since it enables meaning making of representations. Any pupil studying science needs to develop fluency in visualisation or 'metavisualisation'. The key aspect of metavisualisation is the ability to visualise (make meaning of) a representation in the different special dimensions it may occur (Gilbert 2008). Although visualisations are most often visual, they can convey information by using other sensory modalities, such as sound, smell and touch. Culture plays an important role, in the sense that mental models can be expressed and mediated through the cultural tool of drawing. During learning we habitually use external representations surrounding us to construct internal representations in our minds. However, we have no direct evidence of the existence of internal representations. We cannot physically manipulate mental representations to evaluate their validity. Often we are called to convert our mental representations into external presentations, such as during communication for example when writing a scientific paper or composing an email (Gilbert 2008).

Children's learning about animals may be investigated by examining the mental models revealed through their talk and drawing when they come face to face with live or preserved animals. The mental model is the person's personal knowledge of the phenomenon. This knowledge will in certain aspects bear similarities and in others differences to scientifically accepted knowledge, which in the case of this chapter is the appearance of the organisms and their ecological habitat (Reiss and Tunnicliffe 1999). The representations may be written descriptions, verbal descriptions, drawings or three-dimensional models. In this context, observation emerges as an essential skill for scientific learning, which is here understood to mean active looking in search of understanding (Tomkins and Tunnicliffe 2006). Studies on child's internal model have taught us not to assume that children's drawings are print-outs of the internal representations that underpin the topics drawn (Jolley 2010). Although children can pick up ideas from each other in the intimate situation at their benches, children never 'merely copy' and that meaning-making is always a transformative process even when copying (Hopperstad 2010; Kress 1997).

Information in the brain is processed on pre-existing 'schemata' or mental knowledge maps. New knowledge can either be integrated into existing schemata, as Piaget termed it 'assimilation', or the schemata are reorganised to adapt the new information or 'accommodation' (Smidt 2011). Each person processes new matter uniquely as individuals according to their mental knowledge maps influenced by their cultural and biographical experiences.

9.4 Interpreting a Diorama

Individual drawings by children tend to vary appreciably. Analysing drawings by the same child may reveal idiosyncrasies that illuminate the influences on how the animals and plants in the dioramas were interpreted and visualised. In class, the mental model expressed in drawing is mainly from imagination and basically influenced by knowledge held by the pupil at that moment. At this stage, children mainly draw in the iconic mode being able to make mental images of objects and do not need to experience the object physically. Bruner (1966) termed this the *Iconic Stage,* where information is stored visually in the form of images (a mental picture in the mind's eye). Children (5–8 year olds) produce different drawings when they draw from imagination compared to when they copy an object (Gardner 1980).

At the museum, children mainly drew from imagination, but also party influenced by previous knowledge and now also the novelty factor of the unfamiliar museum. Iconic mode is still predominant at this point. The evidence from class and pre-diorama drawings (museum) are indicative of *Intellectual Realism* as coined by Luquet and Piaget. The mental model is still mainly formed by 'what the child knows.' The diorama drawings are now the result of observation and also in some cases imagination. Drawings are still mainly iconic, but they increasingly show organisms in context. Students show a greater degree of *Visual Realism* here, drawing things they 'see' and representing these as they occur. However, most students still resort to their iconic forms to show what they saw thus operating from *Intellectual Realism*. Arnheim (1974) suggested that a child will draw an object which will show the defining features (as the child see's them) in the simplest way for the child to be able to draw them within a piece of paper (2D space).

This would suggest that the *Representational Stage* has been reached by most pupils, that is, when the child makes basic and generalised representations of organisms. The human figure consists of a round form, inner shapes that become the eyes and arms as two lines radiating from the circle. The child draws just a "dog" rather than his or her dog (Gardner 1980; Kellogg 1970; Lowenfeld 1963; Striker 2001). Apart from the principle of differentiation that applies broadly across a wide range of tasks and subject matter, a second principle becomes evident. This is the desire to create a likeness to the object. This desire to capture the object and represent it truthfully guides the direction the differentiation of form takes.

When they were asked to draw in school and pre-diorama, children did so from imagination and previous knowledge while it was increasingly from observation during the diorama task. Most class and pre-diorama drawings were complete constructions of scenes created in the minds of the children, such as gardens, forests and beaches. The diorama drawings were done from looking at particular settings, and so show a personalised representation of the preferred setting. This left less room for creativity, and rather drawings showing varying degrees of resemblance to the dioramas. However, almost all diorama drawings show modifications from the actual, where the pupils give their personal 'touch' to the drawing by selecting items to represent from the setting and adding others from their memories.

9.5 The Interpretation Model

A few major and well-accepted interpretation models commonly encountered in science education and informal learning literature include: (a) Contextual Learning Model (Falk and Dierking 2000), (b) Acuity Model, (c) Model Based Learning (Buckley and Boulter 2000) and (d) Activity Theory (Engeström et al. 1999).

Activity System (Fig. 9.1) that originates from Activity Theory (Engeström et al. 1999) is more applicable to cultural tools and so more appropriate and adaptable in the case of interpreting a museum exhibit such as a habitat diorama. The new Diorama Interpretation Model is derived from Activity System, which has been adapted to become applicable to museum exhibits with the inclusion of additional features as suggested by empirical evidence from my research.

Figure 9.2 above, presents the Interpretation model, which presents six interrelated factors as explained in Table 9.1. Focus, Artefact, Group and Subject emerge

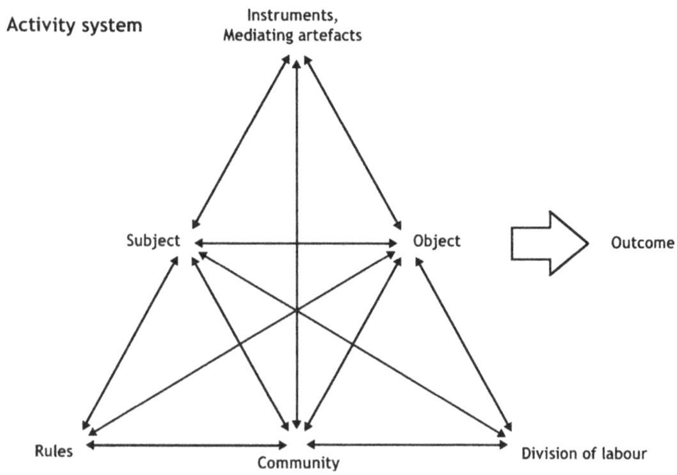

Fig. 9.1 Activity system (Engeström et al. 1999)

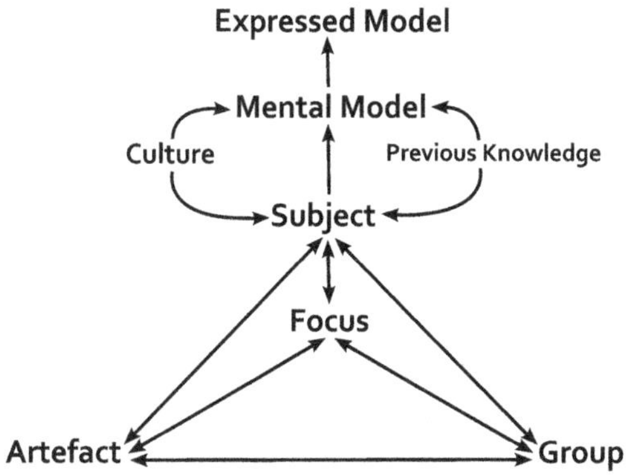

Fig. 9.2 Interpretation model for cultural tools

Table 9.1 Interpretation model terms defined

Focus	The idea, topic or location represented by the artefact and of interest to the subject and/or group, such as 'habitat'
Subject	The person observing the diorama, i.e. student or visitor
Artefact	The mediating tool; a diorama, picture, 3D model or other media forms
Group	The group of people i.e. friend or family, with whom the subject experiences the artefact
Mental model	The personal representation of the artefact or focus held in the subject's mind
Expressed model	The external representation of the mental model
Culture	The sociocultural imprint of the family, country and society
Previous knowledge	What the subject already knows about the focus

from Activity System, while Culture, Previous Knowledge, Mental Model and Expressed Model emerge from data (Mifsud 2015). This model may be used to interpret museum objects or artefacts.

The *focus* here refers to a habitat or a natural object rather than objectiveness of the reality, which for Leont'ev has social and cultural properties. For the "person-object-relation" (POI) theory the creation of interest needs a situation-specific interaction between person and the object. The *focus* generates situational interest that is important for learning particularly in non-formal learning settings (Scheersoi 2009). Situational interest emerges from the viewing of the diorama, but individual interest is also required and this resides within the individual or the *subject*.

The *Subject* is the individual person engaged with the exhibit. Does he or she view habitat dioramas as representations of natural settings? What is obvious to the expert might not be so to the novice. Primary school children are normally novices to learning from visualisations. There is also the risk of dual representation in that novices may focus attention on the object itself rather than the intended meaning. People may interact with an artefact alone or with others, in the model, the *group*. In the sociocultural context learning occurs while experiencing a museum artefact with other people. Experiences with others occur initially on the inter-mental plane and later individually internalized on the intra-mental plane.

The *artefact* can be the museum exhibit, model or other medium that mediates understanding of the *focus*. The *artefact* generates interest in the *focus* and interest is affected by experiences and personal history (Falk and Dierking 2000). Interest facilitates 'new' knowledge about the *focus* constructed by interacting with the *artefact*. Learners use models to assimilate 'new' information and incorporate this with their 'prior' knowledge about the *focus* to build a new mental model (Buckley and Boulter 2000). The mental model is reinforced and routinely used if judged to be adequate, but revised or rejected if considered to be inadequate. If a habitat diorama qualifies as a museum object it may serve as a model for biological learning. Understanding museum objects is complicated since the categories of meaning are more vague with objects than with texts. Every person interacts uniquely with a museum object to form a unique mental map depending on cultural and prior biographical experiences (Hooper-Greenhill 2000).

In informal settings learners readily make associations between what is already known, *Previous Knowledge,* and new knowledge. Museums allow meaning making by connecting with what is already known and comparing the unfamiliar with the new (Hein 1999; Hooper-Greenhill 2000). Representation frequently tries to make sense of previous experience and is a dynamic, constructive act that actually shapes the experience itself (Matthews 2003).

The role of *Culture* is important to the social constructivist who would also consider the role of peers as visitors interact in groups. Wertsch (1991) does not consider the person as a decontextualized individual, but reasoning is conceived to be an inherently social and cultural process of meaning making. It is interesting to see how children's drawings make explicit their beliefs and attitudes, which are not free from stereotypes and simplifications that exist within the culture of the school (Moussouri 1997). Drawing is a mode that is socially shaped and culturally given resource for meaning making. Drawings from different societies and regions of the world do not support the notion of a universally valid and culture-free mode of expression or *Expressed model* (Golomb 2004).

As each person has their own mental maps of knowledge depending on their prior cultural and biographical experiences, each person will process new matter in ways that are specific to them as individuals (Hooper-Greenhill 2000). The child's personal knowledge of a phenomenon or main features of an object are held in his or her *Mental Model* and when asked to draw (expressed model), the child does so from the internal model (Reiss and Tunnicliffe 1999; Cox 1992).

9.6 Applying the Model

The diorama interpretation model is here applied to data obtained from a nine-year-old boy in the fifth year of primary school (Fig. 9.3). The *subject* in this case is Jeremy, a pupil observing the habitat diorama. The *group* is his class and particularly the group of pupils he was experiencing the diorama with. His preferred

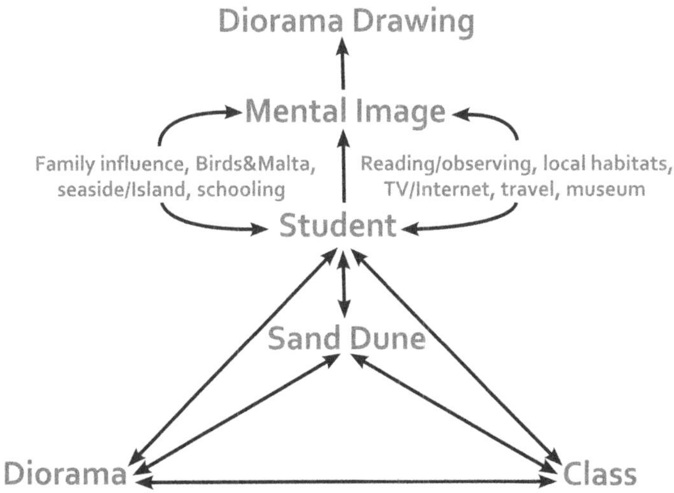

Fig. 9.3 The interpretation model applied (Mifsud 2015)

Fig. 9.4 The sand dune diorama

habitat diorama is the Sand Dune (Fig. 9.4), which is in this case the *artefact*. So, we have a static 3D artefact with various birds, a typical sand dune reed and a very prominent traditional Maltese boat resting on a bed of sand with no pained background. The pupil observes the diorama, which acts as the mediating tool to aid in the interpreting and understanding of the *focus* or the Sand Dune habitat represented by it. He does this in the company of his colleagues, which may influence the way he 'sees' the diorama and what he notices. The role of the more knowledgeable peer may come into play here. However, it is difficult to determine to what extent this occurs and what the actual influences would be.

The observation of the Sand Dune diorama (*artefact*) creates a mental model, which is than expressed as a drawing. This representation is a likeness or simulation of the museum object. In learning we often use a representation (model) to build an internal representation, held in our viewer's mind. Very often, we are called to convert our personal knowledge of an object held in the mental model into an external presentation such as a drawing (Reiss and Tunnicliffe 1999).

There are some similarities, such as the bird on the boat, the flying bird, the two brown birds on the left and the bird on the rock on the right. However, there are differences too, such as the bird opening the mollusc shell (central), the blue background, no reeds drawn and notably the boat facing the other way. These differences are the evidence of *previous knowledge* merging with what was perceived from the diorama. During the interview, Jeremy expressed his interest in wildlife, especially sea life and sea birds. He read about the bird opening mollusc shells and wanted to add seaweed too. The *cultural* influence is noted in the inclusion of blue 'sky' background and the rather standard way the birds were represented, that is, in side view with beaks, both legs, eyes and in aerial view. Golomb's (2004) findings on graphical representation and central positioning of animals, balance and linking of the different features in composition on the drawing are reflected in Jeremy's drawing (Fig. 9.5). The choice of diorama is mainly influenced by the child being an island inhabitant and so well acquainted with the seaside habitats (Mifsud 2015).

Fig. 9.5 Drawing of the sand dune diorama

9.7 Strengths and Limitations

The strengths of the model lie in the manner it links together the elements involved in the interpretation of an artefact (mediating tool) to understand the message it conveys, for example Natural History Dioramas present flora and fauna in their habitat showing possible ecological relationships. It elucidates how a learner may understand a topic as mediated by an artefact to construct an intangible mental model to create a tangible expressed model (a drawing). The interaction with peers, the cultural baggage possessed and knowledge held may influence the mental model constructed. Potentially, this may apply to various topics as presented or modelled by 2D, 3D or virtual mediating tools. This may be done in different learning situations in science and other areas of formal, non-formal and informal settings.

The model's limitations lie in:

(a) Firstly it assumes that the learner would use the artefact affectively. However he might concentrate on specific items in the setting ignoring the bigger picture. An *artefact* may not be effective in understanding the *focus*. Some features may actually distract the learner or capture his attention for aesthetic reasons only.
(b) The degree or quality of interaction between the learner and his peers may be uncertain. The role of a more knowledgeable peer may be difficult to determine.
(c) The *mental model* is very personal and varies from person to person. It is also uncertain how far the mental image is modified and developed by the learning experience.
(d) The *expressed model* is rarely a 'true' replica of the *mental model*. A drawing is normally a selection of what really interests the person from observation and earlier learning.
(e) *Culture* and *previous knowledge* are both long-term factors, which influence the way persons learn, acquire new knowledge and build *mental models*. However, it is difficult to measure the effect these have on the learner and his *mental model*.

9.8 Conclusions

Dioramas are elaborate depictions of constructed habitats that may serve as a model for real habitats and enable visitors to discover and learn about flora and fauna. Natural history dioramas have the potential to act as biological models for visualisation and interpretation of animals and plants. They are meant to promote learning and have a potential role as models in promoting model-based learning.

The new Interpretation Model presents a way to interpret museum settings or applied to other artefacts such as pictures, 3D models and other media forms. It presents the elements involved in making sense of an artefact and the subsequent mental model formed. This is a very personal process, differs from viewer to viewer and can be expressed in various ways such as drawings. Notwithstanding its limitations, the Interpretation Model may be a valuable tool in understanding how museum objects and settings are interpreted and understood.

References

Arnheim, R. (1974). *Art and visual perception: A psychology of the creative eye*. Berkeley: University of California.

Ash, D. (2004). How families use questions at dioramas: Ideas for exhibit design. *Curator, 47*(1), 84–100.

Bartoszeck, A. B., Tunnicliffe, S. D., & Rocha da Silva, B. (2009). *Investigating Brazilian children's concept of insect*. Paper presented at the 8th conference of the European Science Education Research Association, Istanbul, 31 August – 4 September 2009.

Bruner, J. S. (1966). *Toward a theory of instruction*. Cambridge: Harvard University Press.

Bruner, J. S., Goodnow, J. J., & Austin, G. A. (1956). *A study of thinking*. London: Wiley.

Buckley, B. C., & Boulter, C. J. (2000). Investigating the role of representations and expressed models in building mental models. In J. K. Gilbert & C. J. Boulter (Eds.), *Developing models in science education* (pp. 119–135). London: Kluwer Academic Publishers.

Cox, M. V. (1992). *Children's drawings*. London: Penguin Books.

Eberbach, C., & Crowley, K. (2005). From living to virtual: Learning from museum object. *Curator, 48*(3), 317–338.

Engeström, Y., Miettinen, R., & Punamäki, R. L. (1999). *Perspectives of activity theory*. Cambridge: Cambridge University Press.

Falk, J. H., & Dierking, L. D. (2000). *Learning from museums: Visitor experiences and the making of meaning*. New York: AltaMira Press.

Gardner, H. (1980). *Artful scribbles: The significance of children's drawings*. New York: Basic Books Publishing.

Gatt, S., Tunnicliffe, S. D., Borg, K., & Lautier, K. (2007). Young Maltese children's ideas about plants. *Journal of Biological Education, 41*(3), 117–121.

Gilbert, J. K. (2008). Visualization: An emergent field of practice and enquiry in science education. In J. K. Gilbert, M. Reiner, & M. Nakhleh (Eds.), *Visualization: Theory and practice in science education* (Vol. 3, pp. 3–24). Dordrecht: Springer.

Golomb, C. (2004). *The child's creation of a pictorial world* (2nd ed.). Mahwah: Lawrence Erlbaum Associates.

Gordon, I. E. (2004). *Theories of visual perception* (3rd ed.). New York: Psychology Press-Taylor & Francis Group.

Hein, G. E. (1999). The constructivist museum. In E. Hooper-Greenhill (Ed.), *The educational role of the museum* (pp. 73–79). London: Routledge.

Hein, H. S. (2000). *The museum in transition: A philosophical perspective*. Washington: Smithsonian Books.

Hooper-Greenhill, E. (1994). *Museums and their visitors*. London: Routledge.

Hooper-Greenhill, E. (2000). *Museums and the interpretation of visual culture*. London: Routledge.

Hopperstad, M. H. (2010). Studying meaning in children's drawings. *Journal of Early Childhood Literacy, 10*(4), 430–452.

Huxham, M., Welsh, A., Berry, A., & Templeton, S. (2006). Factors influencing primary school children's knowledge of wildlife. *Journal of Biological Education, 41*(1), 9–12.

Insley, J. (2007). Setting the scene. *Museums Journal, 2*(107), 33–35.

Jolley, R. P. (2010). *Children and pictures: Drawing and understanding*. Oxford: Wiley-Blackwell.

Kellogg, R. (1970). *Analyzing children's art*. Palo Alto: National Press.

Kress, G. (1997). *Before writing. Rethinking the paths to literacy*. London: Routledge.

Lowenfeld, V. (1963). *Your child and his art*. New York: Macmillan Publishing Company.

Matthews, J. (2003). *Drawing and painting: Children and visual representations* (2nd ed.). London: Paul Chapman Publishing.

Mifsud, E. (2015). *The visualization of natural history museum habitat dioramas by Maltese school children*. Unpublished PhD thesis, University College London Institute of Education, London.

Moussouri, T. (1997). The use of children's drawings as an evaluation tool in the museum. *Museological Review, 4*, 40–49.

Patrick, P., & Tunnicliffe, S. D. (2011). What plants and animals do early childhood and primary students' name? Where do they see them? *Journal of Science Education and Technology, 20*(5), 630–642.

Paddon, H. (2009). The important role of natural history dioramas in biological learning: Curatorial responses to natural history dioramas. In S. D. Tunnicliffe & A. Scheersoi (Eds.), The important role of natural history dioramas in biological learning. *ICOM Natural History Committee Newsletter*, 29, 1–40.

Peart, B., & Kool, R. (1988). Analysis of a natural history exhibit: Are dioramas the answer? *The International Journal of Museum Management and Curatorship, 7*, 117–128.

Pipueras, J., Hamsa, M. K., & Edvall, S. (2008). The practical epistemologies in the museum: A study of students' learning in encounters with dioramas. *Journal of Museum Education, 33*(2), 153–164.

Rapp, D. N., & Kurby, C. A. (2008). The 'Ins' and 'Outs' of learning: Internal representations and external visualizations. In J. K. Gilbert (Ed.), *Visualization: Theory and practice in science education* (Vol. 3, pp. 29–52). London: Springer.

Reiss, M. J., & Tunnicliffe, S. D. (1999). Conceptual development. *Journal of Biological Education, 34*(1), 13–16.

Reiss, M. J., & Tunnicliffe, S. D. (2007). *Opportunities for learning in biology*. Paper presented at the NARST annual international conference, New Orleans, 15–17 April 2007.

Scheersoi, A. (2009). The important role of natural history dioramas in biological learning: Biological interest development at natural history dioramas. In S.D. Tunnicliffe & A. Scheersoi (Eds.), The important role of natural history dioramas in biological learning. *ICOM Natural History Committee Newsletter* 29, 1–40.

Smidt, S. (2011). *Introducing Bruner: A guide for practitioners and students in early years education*. London: Routledge.

Stern, T. (2009). An afternoon among dioramas at Yale Peabody Museum. In S. D. Tunnicliffe & A. Scheersoi (Eds), The important role of natural history dioramas in biological learning. *ICOM Natural History Committee Newsletter*, 29, 1–40.

Striker, S. (2001). *Young at art*. New York: Henry Holt and Company.

Tomkins, S., & Tunnicliffe, S. D. (2006). Bring back the nature table! *Environmental Education, 82*, 8–11.

Trowbridge, J., & Mintzes, J. J. (1985). Student's alternative conceptions of animals and animal classification. *School Science and Mathematics, 85*, 304–316.

Trowbridge, J., & Mintzes, J. J. (1988). Alternative conceptions in animal classification: A cross-age study. *Journal of Research in Science Teaching, 25*(7), 547–571.

Tunnicliffe, S. D. (2002). The educational value of natural history collections in learning about biodiversity. *The Biology Curator, 22*, 27–40.

Tunnicliffe, S. D. (2005). What do dioramas tell visitors? A study of the history of wildlife diorama at the museum of Scotland. *Current Trends in Audience Research and Evaluation, 18*, 23–31.

Tunnicliffe, S. D. (2013). Editorial: Animals and plants in natural history dioramas in museums: Specimens or objects. *Journal of Biological Education, 47*(4), 189–191.
Tunnicliffe, S. D., Gatt, S., Agius, C., & Pizzuto, S. A. (2008). Animals in the lives of young Maltese children. *Eurasia Journal of Mathematics, Science & Technology Education, 4*(3), 215–221.
Wertsch, J. V. (1991). *Voices of the mind: A sociocultural approach to mediated action*. Cambridge, MA: Harvard University Press.

Edward Mifsud obtained his doctorate at the Institute of Education at the University of London. His main research interests are primary science, environmental education, museum learning and learning in informal settings. Edward started his teaching career in 1997 at the Liceo where he taught general science and Chemistry at all the levels of the secondary school. In 2009, Edward started teaching A-level Biology at the University of Malta Junior College where he was recently promoted to Senior Lecturer. He is also member on the EU scholarship awarding panel MasterIt and member on the National Education for Sustainable Development Strategy (NESDS) board.

Part III
Science Learning Activities Involving Dioramas

Chapter 10
Constructing *and* Reviewing Dioramas: Supporting Beginning Teachers to Think About Their Use to Help Children Understand the Work of Natural History Scientists

Debra McGregor and Jennifer Gadd

10.1 Introduction

Dioramas have traditionally been used to present realistic glass-cased representations of a range of habitats and the flora and fauna that naturalistically co-exist there (Tunniciffe and Scheersoi 2015). These kinds of three-dimensional models have provided static exhibits typically found in museums aiming to show-case for visitors illustrations of the specimens that normally inhabit particular ecological contexts. These reconstructed environments can provide miniature (or even life-sized) representations of typical landscapes and their natural inhabitants. Classically these dioramas have included a background painted to provide perspective and a context as well as preserved animal and plant specimens (Reiss and Tunnicliffe 2011). Reiss and Tunnicliffe (2011) discuss how varied the scenes can be (from an Angolan savannah with impala and wildebeest in it, to wild Alaskan landscapes containing mythical creatures such as Valhalla and even a Kenyan watering hole with grazing giraffes) to illustrate for visitors' narratives or stories about wildlife, with a view to offering opportunities for learning through interpretation by viewers. Piqueras et al. (2012) discuss further how dioramas have also been constructed to purposely present intrigue for viewers within a display. They describe a diorama with a female white-tailed eagle eating the remains of a roe deer in a snowy landscape. In the distance there are a pair of hooded crows (Piqueras et al. 2012, 81) watching and waiting, perhaps considering ways to steal food from the eagle! There are also fox footprints in the artificial snow and the deer's head is missing (apparently is it common place for foxes to take away the heads of the animals they have hunted). The aim of this 'scene' is to promote visitors' curiosity about the relationships between the animals in the snowy setting and to wonder about what has happened and what might unfold

D. McGregor (✉) · J. Gadd
Oxford Brookes University, Oxford, UK
e-mail: dmcgregor@brookes.ac.uk

A. Scheersoi, S. D. Tunnicliffe (eds.), *Natural History Dioramas – Traditional Exhibits for Current Educational Themes*,
https://doi.org/10.1007/978-3-030-00175-9_10

next. The aim of the diorama displays with beginning teachers, in the study reported here, was to raise their awareness and appreciation of the potential use of dioramas to teach about natural history. It was hoped that the students would engage in reflecting about how *constructing* and *reviewing* dioramas developed *understanding* about natural history. Some of the dioramas the students produced, did indeed offer opportunities to wonder about (and deliberate over) the story being told through the displays. It was assumed that through working with others in a socially constructivist manner, toward the common purpose of generating a series of home-made dioramas presented in a museum-like display, would provide a mutually useful interactive learning experience for everyone (Falk and Dierking 2000; Falk et al. 2004). The objective was to invite each other to view their home-made dioramas to cogitate on whose natural history work might be portrayed in each shoe-box model. The intent was also to engage the beginning teachers in reflecting on their experience of constructing and inspecting dioramas to consider how they might be useful in future for (their own) teaching. There was also an attempt to explore whether the processes of making *and* reviewing could extend Tunnicliffe and Scheersoi's (2010) four-stage response, cited in Rennie 2014 (p.125), of "identify – interest – interpret – investigate" to dioramas. The evaluative approach of the workshop-like session explored, as Jakobson and Davidson (2012) suggest how the production of artefacts (the student-constructed dioramas) contributed to, and mediated learning.

10.2 The Approach

The project reported on is a piece of evaluative action-research carried out by two tutors at Oxford Brookes University. It was undertaken to explore what (and how) beginning teachers could learn about natural history through making dioramas. The interventional experience, described in more detail later was designed as an active collaborative workshop session whereby the students co-constructed home-made dioramas in triads. On-going conversations, reflective discussion (and responses to an evaluative questionnaire at the end of the session) suggested how dioramas were helpful in learning about natural history and how they might be more effectively utilised to support learning in future.

10.3 Focusing the Diorama-Making on Content Required by English Schools

Beginning primary school teachers in England learn to teach 5-year-old to 11-year-old children. Whilst training they need to develop a range of pedagogical skills to deliver a high-quality education whilst also demonstrating the subject knowledge required to teach 11 different subjects in the Department for Education's Primary National Curriculum (2013). These 11 subjects include science.

The science curriculum is intended to allow children to 'develop a sense of excitement and curiosity about natural phenomena' (Department for Education 2013, 144). It is divided up into year groups and then into different biology, chemistry and physics topics. Each year group has at least one topic related to natural history; either plant biology, human biology or animal biology.

The national curriculum has 'statutory requirements' which must be taught, but also a 'notes' section with a suggestion for lesson ideas and objectives. For each topic a famous scientist is recommended as one to study. For the natural history topics, these could include David Attenborough, Jane Goodall, Mary Anning, Rachel Carson and Charles Darwin. In addition, an objective for pupils in year six (aged 11 and 12 years old) is to be able to 'talk about how scientific ideas have developed over time' (2013, 166). These beginning teachers therefore needed to be familiar with each of these scientists and their work in order to teach children about them effectively. This workshop was thus devised to explore how classroom dioramas support and mediate teaching this aspect of the curriculum.

10.4 The Students Involved in the Project

Oxford Brookes University trains approximately 600 students a year to become primary school teachers teaching 5-year-old up to 11-year-old children. There are many different training routes including a 1 year long course for post-graduates, and a three-year long course for undergraduates. Regardless of the route taken into teaching, these 'beginning teachers' are given workshops, seminars and lectures in all aspects of the national curriculum by subject experts in order to ensure that they have both the pedagogical skills and subject knowledge required to teach.

For the purposes of this action research project the second-year cohort of beginning teachers were invited to be involved. The cohort consisted of 71 students aged between 19 and 35 years old. All of the beginning teachers had experienced at least 12 weeks of working in a primary school and were in the process of preparing for a further 8-week long placement. As a result they had some experience in working with young children and were developing their pedagogic expertise and subject knowledge. On their placements, the majority of students had taught one science lesson each on any topic. Their reflective comments, therefore, are based on a brief, initial experience as a beginning teacher in school.

10.5 The Diorama Workshop Session

In the second year of the Batchelor of Arts (Educational Studies) degree course, the beginning teachers are engaged in 18 workshops or seminars lasting one and a half hours per week. Each of these seminars focuses on a different aspect of subject knowledge. For the purposes of this evaluative project, a new seminar was designed. The intention of this seminar was two-fold. Firstly, the main intention of the seminar was to introduce the beginning teachers to the scientific contributions and personal stories of the natural history scientists named in the 'notes' section of the national curriculum; David Attenborough, Jane Goodall, Mary Anning, Rachel Carson and Charles Darwin. Secondly, the seminar aimed to allow beginning teachers to experience, in role as primary aged children, the process of constructing (and examining) a diorama and to consider the benefits of conducting a similar activity, through creating a museum-like display in a classroom in school.

To assess the student's familiarity with dioramas they were asked about their views and previous experience of them. The 71 beginning teachers were split into 3 (successive) teaching groups organised to work in (mostly) triads to work on their diorama projects together. Each trio was allocated a card with the name of a scientist written on. These were distributed at random. The students were requested not to divulge which scientist they had been allocated; they were invited to 'reveal' their scientist through their displays. Each group of three were given a shoe box within which to create a diorama, and a range of everyday modelling materials such as paper, card, tissue, coloured cellophane, paints, hot glue guns and cutting knives.

Before beginning to make the dioramas, the beginning teachers were invited to create their own success criteria for their displays. They were asked to consider what they might expect a primary school child to achieve during the activity. It was determined that, in order to be successful, the following would need to be achieved during the activity:

- The diorama should depict the scientist and their contribution to the field of natural history including what they looked like and what they achieved.
- Other students should be able to work-out which scientist was depicted by looking at the diorama.
- The diorama should be accompanied by a 'fact sheet' of key ideas about the scientist's life and work (a cultural convention noted by Achiam et al. (2014, 4) to relate to a real museum exhibit).
- The students should work collaboratively in their group, taking turns and negotiating when making decisions about what or how to make something.
- The diorama needed to be fully complete within an hour of creation time.

During the creation of the dioramas the beginning teachers were provided with a tablet computer connected to the internet in order to research their scientist fully. They were encouraged to find out as much as they could about the scientist's personal life and their contribution to the field of natural history. The tutors moved from group to group collecting field notes and photographs (see Figs. 10.1, 10.2, 10.3, 10.4, 10.5a, 10.5b, 10.5c, 10.5d) to document the artefact-construction-process. Once the building-dioramas phase was complete the beginning teachers then took time to visit the other dioramas as if at a local museum reviewing the display boxes. Each group jointly construed their explanatory narratives about which Natural History scientist was portrayed and then verified their predictions through uncovering and reading the accompanying fact sheet. Utilising the success criteria they also considered how well each diorama achieved the display aims. Photographs of some of the finished artefacts were taken (see Fig. 10.5a, 10.5b, 10.5c, 10.5d).

Fig. 10.1 The students begin by painting their shoe boxes in an appropriate colour

Fig. 10.2 This 'David Attenborough' diorama has blue acetate representing the sky

Finally, the students were invited to complete a short questionnaire. They were asked to consider how they would describe a diorama; what they thought they learned through making a diorama; what they thought they learned through reviewing others' dioramas; which areas of science lent themselves to-be-learnt-about through making-your-own-diorama; what they thought children might benefit from through dioramas and whether there were particular stages in learning where they were most helpful. The students were also asked about their understanding of the focus of the dioramas, the natural history scientists.

Fig. 10.3 Some groups went outside the classroom to collect leaves, sticks and pebbles to add to their dioramas

Fig. 10.4 Beginning teachers had to research their scientist carefully before representing them in the diorama

Fig. 10.5a A final Charles Darwin diorama focused on the origin of species

Fig. 10.5b A final Mary Anning diorama

Fig. 10.5c A final Jane Goodall diorama

Fig. 10.5d A final Rachel Carson diorama

10.6 Findings

10.6.1 Views of Dioramas Before Being Introduced to Them

Prior to the diorama session, there were quite a variety of views about the possible nature of dioramas, and 24%, nearly a quarter of the students had not heard of them, reflecting Tunnicliffe and Scheersoi's (2015) view that these kinds of museum exhibits are little used and appreciated for their potential contribution to science education. Most of the students (who had heard of dioramas) considered them to be some kind of 'mini-stage' or 'miniature scene' illustrating a particular landscape or vista of some kind. Few specifically mentioned 'habitats'.

10.6.2 Post Session Views of the Nature of a Diorama

After the experience of making their own and reviewing others' unique constructions, the majority of the students defined dioramas as 3D models of something offering a visual representation of a person, event, moment-in-time or a scene of some kind. During the session re-cycled shoe boxes (see Figs. 10.1, 10.2, 10.3, 10.4, 10.5a, 10.5b, 10.5c, 10.5d) were used to create the in-expensive dioramas. So although the students did not engage in constructing dioramas that were traditionally assembled in open or sealed transparent glass cases they certainly recognised how the opportunity to craft a mini-scene portrayal of the story about a natural history scientist's work, provided a creative and enjoyable learning experience. They also recognised how a form of reality was depicted within them and even though home-made (and not constructed to-scale) dioramas could provide an effective display to aid viewers memorable understanding of something. They recognised that it was possible to fashion dioramas to convey a range of aspects related to natural history including habitats, animals, plants, seasons, life cycles and even evolution.

10.6.3 Students' Views of Their Learning from Engaging in Making *the Dioramas?*

Learning About Dioramas Through Collaboratively Constructing Them When reporting on collaboratively constructing the dioramas, the students indicated it was fun and suggested the 3D displays were surprisingly easy to make! They also thought it was an effective and useful way to learn and generally a successful and engaging experience. In the process of construction the kinds of skills that they suggested were required included using colour (see Fig. 10.1) to convey a context, such as the woodland (see Fig. 10.3), sky (Fig. 10.2), beach, under-the-sea (see Fig. 10.5b, 10.5d) or jungle (see Fig. 10.5c). Creating a 3D space or sufficient area to be able to generate depth (through layers) in the dioramas meant that some triads chose to utilise the boxes in different ways. One group generated much more depth of field by adopting the lid as part of the base, another group wanted to present their 'scene' as a plan view rather than from a side elevation (Fig. 10.5a, 10.5b, 10.5c, 10.5d). Another group literally created a 'window' to peer into a moment-in-time of the life of a scientist.

Learning About Natural History Scientists, Their Life and Work 79% of the students recognised, that through the preparatory processes required to build the shoe-box diorama artefact, they encountered and made sense of significantly more information about the scientists than they had previously been aware of. In their joint endeavours to design and plan the construction of the dioramas most of the students indicated they learned much more about the lives and work of Darwin, Attenborough, Goodall, Anning and Carson.

Working together collaboratively, to make a display, more than just 'arrests attention' (Reiss and Tunnicliffe 2011, 447) of learners, it can offer direct affordance by engaging them in transformative thinking about the object of focus for would-be-viewers. As Rounds (2004) indicates selecting a particular element or aspect (in this study of scientists' lives) to pique curiosity can draw attention of visitors to an exhibit. The to-and-fro discussion, between learners, deliberating over how they will construct a particular depiction (with a variety of everyday materials) promotes consideration of the narrative they want to communicate about the key work of the scientist. Co-constructing with others, a jointly agreed design of aspects of reality from the scientist's work can involve in-depth reflection of ways to represent some kind of phenomena related to natural history.

Some students noted that making the shoe-box dioramas was time consuming. This could be due to the challenge to decide how to depict concepts in an accurate and precise manner within a tight time frame. As Cole (1996) suggests, artefacts, such as the shoe-box dioramas are created and developed during the interactions between the collaborating individuals. The dialectic exchanges involving the to-and-fro considerations regarding the 'make-do' materials rather than the taxidermist's real preserved specimens (Reiss and Tunnicliffe 2011) that would be more readily available for museums to create their collections.

10.6.4 Learning Through Constructing a Diorama

The specific learning processes that the students indicated they felt they engaged in, included researching information (related to Fig. 10.4); transforming or re-presenting key information and modelling in miniature or at least somewhat in proportion so the representations can be accurately interpreted. Prioritising what should be the focus of the Darwin (see Fig. 10.5a), Attenborough, Goodall (see Fig. 10.5c), Anning (see Fig. 10.5b) or Carson (see Fig. 10.5d) dioramas promotes criticality reviewing extensive information to select salient facts to include in a display. To synthesize, collate and simply summarise to represent *observing primates* (Goodall) or *the process of evolution* (Darwin) or *recognising the causal relationship of the slow deterioration of something in the environment as a consequence of an accruing pollutants over time* (Carson), for example, are challenging concepts to convey. Examples of the ways in which narratives about the scientists and their work were understood through the participatory process of making a diorama include one student, for example, developing his appreciation of Darwin from "He discovered a lot of things" to knowing that he was born in Shrewsbury, and he travelled to the Galapagos Islands, where he carried out much of his work and developed his theory of evolution. Another student before the activity indicated he knew Darwin was "British", but after the diorama session he knew more specifically that this scientist developed the theory of evolution through his work on finches on the Galapagos Islands. Other additional information students garnered included details about Darwin's publications and the specificity of his work that evidenced natural selection through looking at finches' beaks and the ways they were adapted to collect berries, eat prickly cacti or insects, for example. The final dioramas produced (see Fig. 10.5a, 10.5b, 10.5c, 10.5d) illustrate how the groups focused on differing aspects of the Natural History scientists work to illustrate. Figure 10.5a, for example, highlights how Darwin's Origin of Species included a description of the evolution of man from primates; another diorama highlighted more specifically work, symbolised with an enlarged magnifying lens, carried out on beaks of finches in the Galapagos Islands and a third one clearly recognised the variation in the general morphology of the finches. Each of these three different examples of home-made shoe-box dioramas provided clear indications of the ways that similar information is 'storied' quite differently by the students. This illustrates Vygotsky's (1978) argument that thoughts are created and conveyed through using cultural tools (such as everyday materials in this case). The reified artefacts (the dioramas) that emerged varied because they were produced through the dialectical relationships involving the minds and socio-historical experiences informing the contributions from the students. The variation of interpretations can enrich the reflective discussion and subsequently what is understood about Darwin's natural history work. The students indicated they learned more about David Attenborough and his work too through constructing the dioramas. Many students recognised prior to making dioramas that he was a presenter of natural history on television and that he made documentaries about animals, but until they engaged in researching his work they did not realise the geographical and biological extent of his natural history knowledge and expertise. The dioramas that depicted his life and work illustrated different contexts (watering hole, jungle and polar regions respectively) and facets of his work (openly observing, hiding unseen and requiring binoculars to view animals in situ in their habitat, quotations about 'penguins as the warriors of the polar regions'). It seemed that creating dioramas, in this way, offered the beginning teachers (as learners) the space and opportunity to generate interpretative narratives (Achiam et al. 2014).

The students knew very little about Jane Goodall prior to the diorama session (in fact only nine students were familiar with her name), but by the time the session had finished many had developed understanding about her work, as a primatologist, from just observing gorillas, to recognising how and when she studied a wide range of apes (including chimpanzees, gorillas and monkeys) in Tanzania. Three dioramas depicted her work in contrasting ways. The first provided a window into the habitat within which she studied chimpanzees symbolised by the brown monkey type shapes with oversized bananas!; the second included a representation of a jungle-like habitat with a group of gorillas in a natural setting, where there was a hut-like building from where Goodall might have lived and observed the animals; the final diorama appeared to represent Goodall as an expert on gorillas explaining their behaviour.

The dioramas of Mary Anning included illustrating her at the beach with her pet dog where there were shells (depicting fossils); the second one provided additional portrayals of an ichthyosaur and an indication of its location in a coastal cliff; the third one just focussed on the sea, beach and cliff-side habitats. Students indicated a five-fold increase in their understanding of her somewhat scavenging life-style on the Lyme Regis coastline through using these scenic displays.

The final scientist considered in the diorama activity was Rachel Carson. None of the students had heard of her prior to the diorama activity. The shoe-box display of Carson's work emphasized the under-the-sea habitat she was concerned about, the creatures that inhabit it and the trash that pollutes it. From knowing nothing about this scientist the students came to realise she possessed keen observational skills informing her hypothesis about the causal relationship between pollution and the detriment of organism's reproductive capacity.

10.6.5 Learning About the Natural History Scientists

Students reporting what they learned about the various scientists through engaging in constructing their group displays and viewing others' dioramas are shown in Table 10.1. Interestingly, they indicated that there was much more impact when they were considering (making *and* reviewing representations of) people they knew little about or had not encountered before (e.g. Goodall, Anning and Carson). It appeared that using the diorama activity to develop understanding and familiarity with the

Table 10.1 Beginning teachers' views of perceived affordance of making and reviewing co-constructed dioramas (as judged through the number of comments)

	Comments prior to the activity	Comments after diorama construction and review
Darwin	91	79
Attenborough	82	89
Goodall	9	52
Anning	7	38
Carson	0	15

work of Darwin and Attenborough, scientists they already knew something about, did not promote quite so much impact on learning about their lives and natural history work.

Although Reiss and Tunnicliffe (2011) emphasize how exhibits are used to relay particular stories in museums, having (teachers and) children create their own dioramas offers learners more affordance in a variety of ways. Arguably, challenging the students to create and construct their own dioramas, offered them a 'real' and 'material' design affordance (Norman 1999) that was physical, in the sense that they handled the resources that they used to communicate something with. There will inevitably be a difference in the way that the same display purposely designed to portray the work of Jane Goodall (Fig. 10.5c), for example, will be interpreted differently by viewers. Those looking at the same diorama may assign quite different meanings to the presence of the construct that could be a house, hut or hide in the shoe-box. It is anticipating the variation of interpretations that generate different stories (perhaps Jane lived there permanently or used it occasionally during observational episodes as a hide when the gorillas were close by or perhaps she only used it as a safe haven in cases of emergency) that may have caused some students longer deliberations over how to tell a particular story through their displays. In a socially constructivist sense, we hoped that the interactive learning processes would serve to mediate each other's thinking and understanding about how to portray aspects of natural history science in differing ways. Through encouraging collaborative working together we anticipated that the students would solve the practical problems as they arose in the diorama construction and in so doing they would dialogically exchange understandings to clarify and make sense of the intention of the group display. In the tradition of social constructivism we assumed that the students would .... 'solve practical tasks with the help of their speech, as well as with their eyes and hands' (Vygotsky 1978, 26).

Engagement in the collaborative production of the dioramas with everyday materials offered a physical affordance (Norman 1999). This promoted much discussion and reflective negotiation about the possible ways that the displays could be constructed. The in-depth thinking and creativity that emerged from these purposeful dialogues was specifically recollected by the students.

10.6.6 Students' Views of Learning from Reviewing *Others' Dioramas?*

Learning About Natural History Scientists, Their Life and Work It is a very similar proportion, around 80% (see Table 10.2), of the students who indicate they learnt about the scientists their life and work, through the interactive processes inherent in reviewing each other's dioramas. Interestingly this is not significantly different from the perceived processing of information required to *make* DIY (Chu 2014) dioramas! It was anticipated that perhaps the *making* of a diorama might involve generally more 'talking', 'thinking' and 'collective decision-making' and

Table 10.2 Students views about the learning processes inherent in making and reviewing dioramas (n = 71)

	Through the constructing process (%)	Through the reviewing process (%)
Perceived emergent understandings	79	80
Learning processes	76	31
Specific learning processes involving collaboration (discussion, negotiations and decision-making centred around completing task, collective actions contributing to completing task)	47	25
Specific learning processes involving physical co-construction skills	27	4

joint-construction (physically as well as psychologically) about the ways to transform the information concerning natural history scientists into a realistic depiction of some kind.

Contrasting the perceived affordances offered through constructing and reviewing dioramas.

The students' comments about their own emergent understandings when comparing constructing and reviewing dioramas appeared to be interestingly quite similar. However, the stark difference they remarked upon, was the extent to which they felt they were learning when making the dioramas rather than viewing them. They noted that there were a variety of collaborative processes engaged in (including more discussion, more task focus and more physical construction) when involved in designing and building the displays. Negotiation and decision-making were important too, as they had quite a short time within which to produce a shared outcome. The collaborative outcome of their shared endeavours was a physical 3D display depicting the life and work of a natural history scientist. Achiam et al. (2014, 4) suggests that this kind of pragmatic imperative can offer physical-geometric affordance. The sequential experience of making and then reviewing others' dioramas could arguably offer the students cognitive affordances because they were in a related situation (of constructing a display) immediately prior to reflecting on, and inspecting another's display (with a similar purpose).

10.6.7 Beginning Teachers' Views About Why *They Would Use Dioramas*

Most students indicated they thought there were a range of benefits for learners. Suggesting why dioramas were appropriate to use for learning, just about half of the students (49%) recognised a wide variety of learning processes involved in the construction and review processes of making the shoe-box models. They commented

Table 10.3 The students' views about when they might use classroom dioramas as teachers (n = 71)

	Percentage of student responses
Introduction (elicitation)	21
Conclusion (summative assessment)	31
Make boring/dry subject matter fun	14
Researching information	24
(Re)Presenting information	7
Cross-curricular/any subject matter	23

on the way that collaborative learning processes (44%) together were fun (6%) and involved discussions, exchanges of ideas, negotiations about what to include and how to make the displays. They realised that researching (13%) for relevant information and considering ways to use resource materials creatively (7%) to represent relevant ideas (14%) that illustrated (9%) and summarised (3%) understandings all involved critical and in-depth thinking (11%). 7% thought that using dioramas for learning would mean they were a memorable activity that 15% said could promote motor and design skills.

10.6.8 *Beginning Teachers' Views About* When *They Would Use Dioramas (Table 10.3)*

The physical ways that making dioramas offered design and material affordance for learning about more abstract ideas was recognised by the beginning teachers when they suggested that constructing and reviewing others' models could be appropriate for any aspect of learning science. They also recognised how building representative displays at the beginning or end of a topic or theme offered teachers insights into the learners' understandings. The beginning teachers also recognised how co-constructing dioramas, offered differentiation by process and outcome and therefore could be conducted in an inclusive manner for all children to be involved. Achiam et al. (2014) suggest that the use of everyday materials offered design affordance; the collaborative nature of working together offered hidden (cognitive and affective) affordances and the shared objective to produce an outcome for peers to appraise offered a perceived affordance.

10.7 Discussion

There is little research literature discussing the ways that home-made or DIY dioramas can be used to promote learning about natural history. Those studies that have been published focus mostly on examining what visitors learn from looking at professionally constructed displays in museums (Davidson and Jakobsson 2012; Ash 2004). Many studies, therefore, have generally focused on the ways that, often permanent, large glass-cased museum displays can promote learning about habitats, ecosystems and the organisms (plants and animals) that usually live there. This chapter offers a fresh and less traditional (post modern) learning perspective. Utilising the approach of encasing a (smaller) visually appealing shoe-box model of the work and life of a Natural History scientist there are similarities with learning from museum exhibits. This kind of home-made exhibit can open up different kinds of discussion and apprehendability, as Allen (2004) suggests this could place the children in a more comfortable framework from which to be curious. The opportunity to co-construct a material depiction of a scientist's work (and discovery) contributes to social constructivist ways of learning because the experience involves the application of both physical and mental collective processes to create a natural history diorama. In this study the processes could be scaffolded through the steps outlined below.

Dioramas traditionally built for museums are very expensive to construct (Reiss and Tunnicliffe 2011) and obviously require much expertise and time to assemble. This chapter reports on an inexpensive, interactive and rapid re-producible way that teachers and children could engage in similar learning (through stories and narratives) like those offered at museums. Affordance is offered in a wider variety of ways than Achiam et al. (2014) discuss through the materials utilised for physical construction as well as the real affordance or even hidden affordances (that is the mis-match between the diorama-builders' intentions and the viewers' consequent story-making).

Additional learning processes beyond those identified by Tunnicliffe and Scheersoi (2015) can be engaged in when learners co-construct their own dioramas for others to examine and review. Set-up as collaborative endeavours, sharing responsibility for co-constructing dioramas can promote development of research skills, in-depth discussion, constructive creativity, deductive and critical thinking, all of which are contributory means by which teachers can support social constructive learning processes.

10.8 Recommendations

The study reported on in this chapter suggests that a post-modern turn (Hooper-Greenhill 2007) could be adopted to teach using dioramas. Rather than the traditional museum approach to preserving and presenting precious or rare artefacts that can only be viewed and abstractly thought about, the approach offered here suggests a more dynamic modernisation of ways in which objects or events could be re-considered. Although this 'modern' process does not involve visiting the treasured, well preserved and robust dioramas of Victorian England, the learning processes engaged in when 'visiting' and 'viewing' artifacts of cultural and scientific interest can be adopted and adapted for the everyday classroom. The learning can be extended to support meaning making through interpretations of DIY exhibits and even promote comparing and contrasting re-constructed narratives of various (similar) events or depicted realities. The process of co-constructing a shoe-box diorama to contribute to a classroom exhibition and then reviewing them, as through visiting-in-role as if at a museum offers affordance and development of distributed cognition (Achiam et al. 2014). The stages that are key, though, to scaffold and mediate the learners through this enjoyable and effective learning process are:

Step 1 Generate a focus for making dioramas that will contribute to a collection of some kind. In the activities described above the focus was the life and work of Natural History scientists. However, as the students suggested dioramas could feature any subject or event (themes proposed included the earth in space, the inside of the human body, electricity and even inventions of any kind).

Step 2 Identify a particular event, person or object that a group should create a diorama to depict. If the groups do not know what each other are doing, this can augment the engagement in the reviewing stage (by intensifying the intrigue and focussed thinking) when it comes to generating a narrative to explain each shoe-box display.

Step 3 Agree the success criteria for an effective diorama (as listed above). This provides clear objectives for each group to aim for.

Step 4 Encourage collaboration to negotiate understandings, plan, design, collect materials required and co-construction of a DIY diorama that is agreed to depict appropriately a particular concept or related facts and information. Working collaboratively in this way can offer affordance and develop distributed cognition through the active contributions from each member of the group. A specific length of time provides a clear target for the group to work effectively together to complete their part of the display.

Step 5 Create a museum-like display by organising the shoe-box dioramas in a thematic way and ensuring the (initially) covered 'fact' card is completed to accompany each box.

Step 6 Visit each others' dioramas to collaboratively (re)co-construct the particular 'stories' (with the success criteria in mind) behind each visual representation (through each shoe-box display).

Epistemologically, it seems, understanding about natural history scientists and their work can be promoted through the scaffolding of pragmatic experiences of building *and* examining dioramas. There are, of course, limitations to the quality (and undoubtedly detail) of the home-made dioramas. However, the processes of peer-peer mediation through working collaboratively can enhance understanding about the ways that scientific concepts have emerged from scientists' particular ways of working. Supporting reflective discussions, by using questions appropriately (Ash 2004), when viewing DIY dioramas (about scientists' work) to consider 'What can you see?'; 'What could it mean?'; 'How might you interpret it/those?'; Who could it be?' can promote reflective and collaborative considerations. Posing these kinds of questions can encourage dilemma and dialogic exchanges focused on imagining and co-constructing stories from visual depictions (of others' interpretative exhibits). This kind of scaffolded process could be perhaps, summarised as "notice-wonder-interpret-suggest" and could provide a step-by-step scaffold for younger and older learners alike.

10.9 Future Work

This study reviews students' views of the building and reviewing of dioramas and considers what they think, as beginning teachers about how dioramas can be used for learning. However, to explore further, in a socio-cultural manner, the nature of different narratives that might emerge, it would be insightful to capture conversations during the different phases of diorama-making and diorama-viewing. Exploring the ways that learners use stories (Arbor 2011) and language to construct scientific knowledge could inform pedagogy about how to best use dioramas in everyday classrooms. This kind of evidence, complemented by post-activity interviews and discussions (Tunnicliffe 2007) would further illustrate the kinds of affordances that home-made dioramas offer for learning about natural history.

References

Achiam, M., May, M., & Marandino, M. (2014). Affordances and distributed cognition in museum exhibitions. *Museum Management and curatorship, 29*(5), 461–481.

Allen, S. (2004). Designs for learning: Studying science museums exhibits that do more than entertain. *Science Education Supplement, 88*, S17–S33.

Arbor, A. (2011). *Small stories for learning: A sociocultural analysis of children's participation in informal science education*. Unpublished doctoral thesis, Rensselaer Polytechnic Institute, New York

Ash, D. (2004). How families use questions at dioramas: Ideas for exhibit design. *Curator: The Museum Journal, 47*(1), 84–100.

Chu, E. (2014). DIY Dioramas. *Popular Photography, 78*(9), 42–43.

Cole, M. (1996). *Cultural psychology. A once and future discipline*. Harvard: Harvard University Press.

Jakobsson, A. & Davidson, E. (Eds.). (2009). *Understanding interactions at science Centers and museums. Approaching sociocultural perspectives*. Rotterdam: Sense Publishers.

Department for Education. (2013). *National curriculum: Primary national curriculum*. London: Department for Education.

Falk, J. H., & Dierking, L. D. (2000). *Learning from museums*. Lanham: AltaMira Press.

Falk, J. H., Scott, C., & Dierking, L. D. (2004). Interactives and visitor learning. *Curator, 47*(2), 171–198.

Hooper-Greenhill, E. (2007). *Museums and education. Purposes, pedagogy performance*. Abingdon/Oxon: Routledge.

Norman, D. A. (1999). Affordance, conventions and design. *Interactions, 6*(3), 38–43.

Piqueras, J., Wickman, P.-O., & Hamza, K. M. (2012). Student teachers' moment-to-moment reasoning and the development of discursive themes – an analysis of practical epistemologies in a natural history museum exhibit. In E. Davidson & A. Jakobsson (Eds.), *Understanding interactions at science centers and museums. Approaching sociocultural perspectives* (pp. 79–96). Rotterdam: Sense Publishers.

Reiss, M., & Tunnicliffe, S. D. (2011). Dioramas as depictions of reality and opportunities for learning in biology. *Curator The Museum Journal, 54*(4), 447–459.

Rennie, L. J. (2014). Learning science outside of school. In N. Lederman & S. Abell (Eds.), *Handbook of research on science education* (pp. 120–144). New York: Routledge.

Rounds, J. (2004). Strategies for the curiosity-driven museum visitor. *Curator, 47*(4), 389–412.

Tunnicliffe, S. D. (2007). The role of natural history dioramas in science education. *The Informal Learning Review, 87*, 11–19.

Tunnicliffe, S. D., & Scheersoi, A. (2010). Natural history dioramas. Dusty relics or tools for biology learning. In A. Filippoupoliti (Ed.), *Science exhibitions: Communication and evaluation* (pp. 186–217). Edinburgh: Museum ETC.

Tunnicliffe, S. D., & Scheersoi, A. (2015). *Natural history dioramas. History, construction and educational role*. Dordrecht: Springer.

Vygotsky, L. S. (1978). *Mind in society: The development of higher psychological processes*. Cambridge, MA: Harvard University Press.

Deb McGregor is Professor in Education at Oxford Brookes University. She was previously Professor in Learning and Developing Pedagogy (and Assistant Director of the Centre for Development and Research) at the University of Wolverhampton. Deb has held a variety of positions in Higher Education, both in the US and the UK. She has also worked as a science, biology and chemistry teacher in secondary schools in the midlands, and an ICT advisory teacher in Staffordshire LA.

Jennifer Gadd is a member of the Primary Education team at Oxford Brookes University. Before joining Oxford Brookes University, Jennifer worked in several primary schools in Oxfordshire and Worcestershire. She was the science and eco co-ordinator in three different primary schools and specialised in teaching upper KS2 in large inner city schools.

Chapter 11
Dioramas and Teachers: Looking, Thinking, Drawing, and Talking

Cristina A. Trowbridge

11.1 Introduction: What Is Going on in This Picture?

A jaguar surveys a cattle ranch. A snowshoe hare hides from a Canadian lynx in a wintry landscape. Two mountain lions, one whose ears are alert and back legs in a position to pounce, looks out from under an overhanging outcrop; another smaller one, lying down, seems to stare at the visitor. The backdrop is the dry, hot, and hazy environment of the Grand Canyon. *Windows on Nature,* the book by naturalist and artist Steve Quinn (2006) describes what it is like to stand in front of a diorama at the American Museum of Natural History (AMNH). The "windows onto nature" vary depending on species and region, but in the hall, mammals take center stage in the scene. What is happening? The diorama recreates visual phenomena of the natural world. The viewer, looking and thinking, is in the position of a discoverer, rather than a passive recipient of knowledge.

I did not immediately start working with dioramas when I began my position in science-teacher professional development at the American Museum of Natural History - although I had many opportunities to do so. Looking back on this period, I would describe my encounters with the dioramas as somewhat superficial. I would stop with teachers and look for a few minutes or make a point about a diorama or share some fact about how they were made, but I did not look at them closely. I had a vision of engaging teachers with natural history dioramas as a vehicle for facilitating conversations, but I lacked a method.

C. A. Trowbridge (✉)
American Museum of Natural History, New York, USA
e-mail: ctrowbridge@amnh.org

A. Scheersoi, S. D. Tunnicliffe (eds.), *Natural History Dioramas – Traditional Exhibits for Current Educational Themes*,
https://doi.org/10.1007/978-3-030-00175-9_11

11.2 What Do You See that Makes You Say That?

As I walked around the Museum observing teachers and students looking at dioramas, I noticed that, like me, they looked but did not do much more. At times a teacher might comment to a student on what an animal is doing, or a student would say "Is it real?" or "What is it?" providing and seeking "right" or "wrong" answers. Teachers, like me, needed support for how to look, think, talk, and listen with their students.

The emphasis on the visual in teacher education has precedents at the Museum. Since its inception in 1869, the American Museum of Natural History has been entwined with public education. Public education was the impetus "which inspired a group of public spirited men who organized and founded the Museum" (Sherwood 1914). AMNH broke away from the "old museum" paradigm and made its priority service to the public rather than the maintenance of a vault of objects (Coleman 1939).

As early as 1880, Albert Bickmore – a founding member and the Museum's first curator of education - began providing lanternslide lectures for groups of teachers on different natural history topics (Sherwood 1914). Lanternslides are photographic images, sandwiched between two glass plates, projected onto a wall. The lectures were popular, and in 1899 more than 13,525 teachers attended (AMNH Annual Report 1899). Bickmore called his teaching method the "Visual Instruction Method", and it was used to develop other education resources for the public schools (Ramsey 1938). These early pedagogical methods were successful and began to shape the Museum's focus on objects and visual images. In her 2002 book *Wondrous Difference: Cinema, Anthropology and Turn-of-the Century Visual Culture,* Alison Griffiths refers to the lectures as the "visual antecedents", to exhibit, photographs, film, video, and IMAX, which are resources for K-12 education today. The emphasis on learning from direct observation of objects and other visual stimuli, as well as learning from experts (i.e., museum staff) would become hallmarks of the Museum's work with teachers and schools.

The potential of informal learning environments lies in the opportunity for visitors to engage (Hein and Alexander 1998) in observation and conversation (Leinhardt and Knutson 2004). Individuals' interactions with the environment, the objects, and one another are critical in museum learning (Falk and Dierking 1992, Hein and Alexander 1998). Social interactions and conversations foster learning in museums (Leinhardt and Knutson 2004).

I wanted to help teachers harness the potential for interactive learning inherent in the museum setting. The challenge I faced was how to design a learning experience that includes intentional conversation that is inclusive, non-judgmental, and inviting. I began creating simple worksheets for teachers to make observations and inferences about a specific diorama. This approach worked within limits. Brief conversations ignited, but proved difficult to sustain. Teachers easily moved away from the role of learner towards that of teacher. Teacher comments ranged from "I like this but I know that my students would not be able to do it" to "how did you get your job?" Other teachers responded with silence, which I interpreted as indicating shyness, fear, or dislike of the activity. Visitors typically believe that staff holds the knowledge about the artifacts or the exhibits, dismissing their own capacity to make sense of what they see.

In 2009, I met Amy Chase-Gulden who taught me Visual Thinking Strategy (VTS), which would change my way of interacting with the dioramas and in turn change how I worked with teachers and students. It was through VTS that I began to engage, look, and learn *with* teachers. In this article, I hope to illuminate the trajectory of how VTS influenced my work with natural history dioramas in teacher professional development. I will share what the VTS experience has meant for the new science teachers I worked with, as well as their students, and include teachers' sketches of the dioramas as an illustration of the phenomenon known in VTS as the multi-perspective of looking.

Visual Thinking Strategy, an interpretive method, is utilized more in art museums than in natural history museums. In the 1980s and 1990s, researcher and cognitive psychologist Abigail Housen developed her theory on aesthetic development based on interviews with art museum visitors. Housen developed a five-stage developmental model to describe a person's aesthetic experience while viewing art (1999). Housen and colleague and museum educator Phillip Yenawine developed Visual Thinking Strategy to support visitors' capacity to look and engage with the artwork. VTS is used in a group setting while looking at art and is facilitated by three guiding questions: *what is going on in this picture?, what do you see that makes you say that?,* and, *what more can we find*? Although these questions were formulated to facilitate examination of art, they have been applied to other disciplines including science (Yenawine 2013).

In VTS, the facilitator listens and paraphrases comments while pointing to specific areas of the artwork or, in my own work with teachers, the diorama. The facilitator connects ideas to broaden viewers' perspectives of what they are looking at (Yenawine 2013).

The VTS structure is deceptively simple and, when it is facilitated well, it can appear effortless. However, it takes practice for a facilitator to become skilled at paraphrasing participant comments and maintaining a neutral position throughout the discussion. A participant may observe something the facilitator has not noticed before and describe it in detail, and the facilitator's initial impulse may be to say "brilliant or great observation!", but the skilled VTS facilitator does not place judgment on person's observations. The educational power of Visual Thinking Strategy is its capacity to level the playing field in a group of participants: no one is the expert in looking and talking about what one sees. Visual Thinking Strategy creates a facilitating environment (Winnicott 1965) that removes the risk of failure, welcomes plurality of perception, evokes curiosity and invites engagement.

11.3 Incorporating Sketching as Part of the VTS Process

I began my work with the dioramas using a modified version of VTS with experienced teachers in a three-year wildlife forensics teacher professional development program with twenty-two middle and high school teachers. The focus was on how scientists use molecular techniques to identify species that are illegally traded; understand the distribution of threatened animals; and plan for conservation. It included learning about how scientists collect scat to further the study of the populations of big cats. The main activities of the workshop were labs. My colleague Dr. Adriana Aquino and I decided to use the dioramas in the Museum's halls of North American Mammals, Asian Mammals and African Mammals in order to give teachers an opportunity to observe the actual big cats in their habitats and place the science and the conservation efforts in a vivid visual context.

We decided to modify VTS with a sketching component. We did this for several reasons. We knew from experience that teachers enjoyed the dioramas and wanted to know how to use them with students but needed support in engaging students in observation and inference. We also knew some teachers were shy to engage in dialog and did not want to talk about what they were looking at. We thought the sketching would promote looking before talking.

The sketching protocol is brief, but sets the tone and establishes expectations. Teachers engaged in a pre-activity of a four-minute sketch in silence. The prompt was: focus your attention on one aspect of the diorama. Maintain that focus of attention, and sketch what you see. I provided paper, pencil and clipboards, and teachers begin putting pencil to paper. The first few times I did this I was amazed by the variation, detail and thoughtfulness in the drawings. I have worked with teachers for many years in science professional development and had not seen this level of pride in drawing. Who knew so many science teachers had such natural abilities and desire for drawing? Teachers asked if they could keep their drawings. Some took photos and gave the sketches to me. It was teachers' evident pride in their work that prompted me to consider another modification to VTS. After the four-minute sketch, I asked teachers to share their drawings with two other people. I did not emphasize the sketch, but rather invited teachers to show others where in the diorama they had focused their attention and what they had learned by doing this. I initially gave 3 min for these conversations. But, these 3 min were filled with laughter and animation with teachers pointing and looking at what they had sketched in the diorama. There was a synchronicity of positive emotions and talking among the teachers about their drawings, which set the tone for looking together and engaging in VTS. This happened consistently, every time the teachers shared their drawings. The four-minutes of looking and sketching in silence, followed by 3 min of discussion in small groups, helped teachers find a voice to enter VTS. The tendency to stray off topic and the reluctance I had seen in some teachers previously, were absent.

One experienced science teacher wrote:

> VTS – absolutely great concept. Could use it in class using a picture on interactive board. I would have them draw – because I enjoyed that activity. I never draw. It opened up a different portion of my brain.

Initial sketching supported the talking. Conversations were lively, and each person's perspective helped all of us see things that we did not initially notice or might have never noticed. In *Mind in Society*, Vygotsky and Cole (1978) emphasized that interaction with others in the environment and cooperation with peers are essential features for learning. The teachers' interaction and cooperation illustrated the core and the power of VTS. Everyone shares some aspect of what he or she is seeing; everyone learns.

I exhibited the teachers' drawings during professional development, and we debriefed the process and commented on the variations where people had directed their attention. Some teachers had natural abilities and knew techniques for shading and using a line to show depth; they shared these techniques with their peers.

The addition of drawing to VTS is like a visual voice, a trace of an observation that can be seen by others. I became convinced that the modified strategy of VTS - i.e., incorporating sketching and small group discussion - was something that I wanted to continue to use in my work with teachers.

11.4 Facilitation of the VTS Three-Question Conversation

To illustrate - I am providing a snapshot taken from my notes of the VTS process in front of the Alaskan Brown Bear diorama with a group of beginning science teachers. The process begins as teachers assemble clipboards, pencils and paper and enter North American Mammals hall. We gather around the Alaskan Brown Bear diorama. Some teachers sit on the floor. I say, "I am putting you on the clock for four minutes to make a sketch of something that has caught your attention. Please no talking." Teachers begin to look and sketch. Some sketch the bear that is standing, noticing how big the head is; someone is focused on the dead salmon nearby; others are sketching the mountains. Someone is sketching the bear and notices the scars on his nose. I say, "We are going to engage in VTS, which is a strategy that has three questions to support us having a conversation about the diorama." I say the questions, and then ask everyone to take a moment to look, and ask for someone to start us off – "What is going on this diorama?" I described the discussion in notes taken that day:

> Two hands go up. Well, Roger who is a little behind me to the right wants to say something and Ann Marie who is sitting on the floor looking wants to talk. We defer to Ann Marie. She is reserved and it surprises us all that she begins. Ann Marie talks about the bear maybe eating the fish, and killed the fish. I point to the salmon and paraphrase her comments. Roger continues and says that maybe it was the animal in the back, and the reason is that the bite on the fish is small. Others start chiming in, and in between comments I paraphrase and ask the teachers what more can we find? David notices the small animal in the back and is trying to describe his expression. Melissa notices the watermarks from the fish being dragged from the stream/pond. I point to the fish and someone says that he hadn't even noticed the fish. Belinda notices the clouds, and Jamiela defers to her about the different clouds. Kenny talks about the history of the diorama and is wondering if the bear is not looking out at us. I point to the bear and paraphrase some of the comments, and we focus on looking at the bears. Erica and Jamiela describe the stances of the bears, the position of

> being alert. Belinda talks about the time of year and others also chime in. Some mention the grass, the snow, and I am pointing and paraphrasing and trying to link ideas about the snow. Miguel gives a detailed description about the snow line being low and then Belinda says it's spring because the salmon spawn in the spring. Melissa says she has just noticed the bear tracks. Erica mentions that the mountain in the back with the swirl of snow looks like an avalanche, and then Miguel says well maybe that is what that cloud is in the middle of diorama an avalanche happening. I ask him to tell us more - what is evidence for this - and he says maybe that is why the bear is standing and listening, because of the position of the ears.

Given the object of observation - a scene of the natural world - I found that the use of VTS with dioramas allows teachers and students to engage in the process of science, which begins with observations made by using the senses (and extensions of our senses). A museum diorama embodies a moment in nature, and the question "*what is going in this picture/diorama?*" prompts students to make observations and inferences. Typically, the responses alternate between direct observations of what they see (i.e. I see a salmon fish) and inferences from those observations (i.e. Belinda thinks it is spring because the salmon spawns in spring). Since this is a critical aspect of the nature of science - the difference between observation and inference -, the second question, "*what do you see that makes you say that?*" encourages students to support their observations and/or inferences (i.e. maybe that is why the bear is standing and listening: because of the position of the ears). These two questions are at the core of scientific thinking and are essential for understanding the process of science. The third facilitation question, "*what more can we find?*" aims at making participants notice that there is always more to find. This is true of science. Current knowledge is based on the available evidence; thus, science knowledge is intrinsically tentative and iterative, based on new observations and evidence.

However, the difference between the object of observation in art and in science makes the nature of the exchange between facilitator and participants slightly different. Observation of an artwork (i.e. the objective description of what we see) is based on the sense of vision; the inferences or interpretation of what we see is intrinsically framed by openness and subjectivity. It can vary depending on whether the observer is an art critic or a school student, but ultimately both are evaluating a product of the imagination.

Besides the exquisite artistic value of the dioramas, these are meant to represent, as realistically as possible, actual places, animals, plants, skies, etc. They were designed to be consistent with material reality. This creates a new challenge for the facilitator, who, following a VTS protocol, avoids providing evaluative feedback and definitively correcting participants. A student, for example, could misidentify an animal. Should we correct him? One alternative could be to invite the observations of other students, thus eliciting alternative hypotheses. The teacher could also encourage the student to find more about it back in the classroom. VTS can thus foster students' reconsideration of their observations and inferences as well as learning from the group conversation.

11.5 VTS: Supporting New Teachers' Induction

In 2012, my position at AMNH changed from delivery of professional development to self selecting New York City teachers (primarily public school teachers) to working with graduates in teacher induction of AMNH's newly founded Masters of Art in Teaching (MAT) residency program. The goals of New Teacher Induction at AMNH are to develop teachers' abilities to: surface student thinking and plan instruction; to strengthen teachers' use of museum resources and pedagogy; and to develop culturally responsive teachers to accelerate student achievement.

The MAT program includes museum and school residencies. During the first museum residency, pre-service teachers (residents), working alongside Museum educators, facilitate conversations with the public about exhibits in the Cullman Hall of the Universe, the Gottesman Hall of Planet Earth, the Guggenheim Hall of Minerals, the Morgan Memorial Hall of Gems, and the Ross Hall of Meteorites. In the school residency, residents placed in public middle and high schools in the New York metropolitan area develop class field trips to science exhibits and cultural halls at AMNH. In the second Museum residency, participants engage in scientific research in the field and lab. Over the 15-month period, residents become well versed in museum pedagogy and AMNH collections.

In my new role in new teacher induction I wanted to spend time with VTS and dioramas, but I was not sure how this approach would be received by a group of 18 beginning middle and high school earth science teachers who had just spent 15 months in the Museum's MAT Residency program. This cohort of teachers knew each other and the Museum well. I framed the purpose of our visits to the diorama halls by stating that part of our work in induction would be strengthening MAT graduate teachers' use of the exhibits and dioramas for school visits, as well as provide time for teachers to engage as learners. I intuitively felt the cohort could benefit from engagement with the dioramas by learning together and from each other and experiencing their own love of learning as a resource for teaching.

Once a month on Friday evenings, the cohort of teachers meets. One of the benefits of the Museum's new teacher induction is access to the dioramas when the Museum is closed. We usually visit a diorama for about 20 min at the end of the evening. I remember vividly the first meeting in which I suggested that we go to the hall of North American Mammals to do VTS. I had collected drawing paper, pencils and clipboards. I said little about VTS to the teachers in the classroom where we had gathered, preferring instead to engage in front of the diorama. In Fig. 11.1 b, c, d, e) are some of the sketches of the Alaskan Moose diorama. I include these to illustrate the variation in teachers' chosen foci of attention and perspective.

Within a few months of my integration of modified VTS with dioramas in New Teacher Induction, the activity had demonstrated its power to provide new teachers with a valued opportunity to think and reflect on their teaching. Teachers were vocal about appreciating the time with the dioramas. "I appreciated having time to look closely at a diorama and just be." Two of the teachers said that VTS with dioramas was a motivator for attending induction meetings, and they looked forward to the

Fig. 11.1 (**a**) Alaskan Moose diorama (Photo D. Finnin/AMNH); (**b**, **c**, **d**, **e**) teacher's sketches

time sitting on the floor in front of a diorama - looking and drawing and talking. Teachers need to engage in learning experiences that prompt curiosity and conversation in ways that parallel what they can do with their students to foster engagement and inquiry.

One first year teacher working with predominately second language learners in an Earth Science high school class commented that she appreciated having the time during new teacher induction to engage with the dioramas.

> Like meditation, it required focus, but was enjoyable. It was meaningful to pay attention to small details that would normally go overlooked. It created an allotment of time that was reserved for silence and peace. It forced us to have a break from the busy schedule of teaching to relax and focus on detail. Specifically, since induction had a lot of discussion on the challenges of teaching, and was the ending of a stressful workweek. It was a great stress releaser that kept my mind active. I've thought about going to the museum on the weekend to draw some of the dioramas.

But this same teacher did not appreciate the facilitated discussion in VTS.

> I know the purpose in a science sense was to think about the subject and ask questions (why do the clouds look like that? Why does the animal have that feature? Etc.) But I wouldn't focus on questions. Instead, I found the enjoyment in drawing and paying attention to detail with the goal of drawing it, not how it came to be in real life. I personally didn't enjoy the discussions that followed VTS, and would often keep drawing through the conversation not really engaged.

Some of the teachers did not like to draw and felt a little frustrated with this aspect of modified VTS. I started to provide writing prompts for teachers who did not wish to draw, inviting them to write for 8–10 min. The writing prompts were typed out on slips of paper and touched on topics that are specific to new teachers' lives and/or topics that I knew were relevant at school. Here I provide some writing prompts that we used in front of the dioramas: "What is getting in the way for you to plan lessons?" or, "Write about a student that challenges you" or, "Reflect on how you are making lessons culturally relevant to students," or "Write about what you are learning about yourself in your new role as a teacher." Teachers were asked to sit in front of different dioramas and write or draw. The goal here was to have teachers reflect on their teaching and find quietness by sitting in front of a representation of a natural setting. Afterwards teachers were asked to share in small groups and then the entire cohort would circle up to have a whole group discussion. One teacher, in her first year of middle school teaching, preferred to write in front of the dioramas. She often shared her writings, which explicitly described the struggles she was having with establishing her presence in the classroom. I felt that by her giving me the writing she wanted someone else to know her reality. Recently, I asked her to comment on this experience of sitting in front of the diorama and writing.

> It was peaceful and relaxing. The museum was quiet and was a calming place to be and allowed me to think about the week/month/year. It allowed me to reflect in a peaceful setting and to think of ways to use the dioramas with students.

Looking at learning outside of schools to support learning in schools can provide insight on how to support novice teachers and strengthen students' engagement and learning outcomes. Informal science learning environments (ISEs) are a potential resource that can support new science teachers. Early exposure to ISEs broadens teachers' perceptions of the instructional support ISEs offer (Kisiel 2014). The work may reveal alternative educational approaches, which increase engagement with groups of students with limited engagement in science education. There is evidence that informal science environments provide a safe environment for teacher learning (Avraamidou 2014). Avraadimou (2014) and Kisiel (2014) recommend additional research for understanding the interactions of informal science environments on middle and high school teachers classroom teaching.

11.6 What More Can We Find?

The responses of the new teachers to the use of VTS with natural history dioramas, and the impact it appears to have in their own teaching are consistent with the recognition of the potential that work with informal science institutions has for developing reform-minded science teacher identity during teacher preparation programs (Avraamidou, 2014). The first year of teaching can fill new teachers with doubt. Teachers succeed and fail in front of many people (students) all day long. It does not take much for new teachers to question their value and identity or career choice, or to experience themselves as incompetent. The development of a positive teacher identity is significant for teacher retention. How can natural history dioramas help? Here is a first-year teacher commenting on working with her students with VTS and dioramas.

> To start, this activity was a very low maintenance activity. As a new teacher, it was great to run a successful activity that was not strenuous. When bringing large groups of students to a museum, it's important to be well planned and knowledgeable. VTS allows you to take some of the stress away from planning the trips, since the goal is to have students think and construct their own understanding, as opposed to looking for a correct answer.

> ...having the students build on each other's images in order to construct a story ... allows the students to see the dioramas in ways they have never seen them before. Yes, we connect the VTS to the Earth Systems. My students were just starting a unit on geoscience. As such, I had the students look for elements of atmosphere, lithosphere, hydrosphere and biosphere within the diorama. It was very effective for showing how all systems interact.

> It was beautiful to see students thinking and constructing for themselves! No matter the level of the student, each student can participate. Each student had ownership of their work!

A few teachers commented how they worked with art teachers for assistance on viewing the dioramas and modifications were made that maybe only benefited the students who like drawing. Here is a second-year teacher commenting about her visit.

> I took students to the museum this year and collaborated with the art teacher. Students had to find artefacts/objects in dioramas to drawn, with a particular focus on texture. I noticed that most students became restless with this activity after 5 minutes. The only students that were completely engaged were those who enjoyed drawing.

The diorama is a resource, and VTS invites the viewer to look and describe what she sees, which is open and allows for various entry points for people to engage in science. Research from the National Research Council on *Learning Science in Informal Environments* (Bell 2009) suggests that when students from non-dominant groups in science are provided with resources, learning experiences in and out of the classroom can support skills and improve sense making in science.

Natural history dioramas serve as a facilitating environment to support new science teachers' development. Teacher responses to the use of VTS and modified VTS, described in this article, suggest that these strategies have a positive impact on teacher learning and reflection. Two strategies I'd like to explore further in relation to modified VTS is its potential as a contemplative practice and its application in the classroom setting, I am interested in how new science teachers transfer the process of looking and thinking from dioramas to their classroom environment. I think there is still more to find about how the use of dioramas can strengthen teacher reflection and student learning. My goals are to continue to observe, talk and listen with teachers as we interact with the dioramas.

Acknowledgments I would like to express my appreciation to AMNH colleagues, Dr. Adriana Aquino and Donna Sethi for their comments and support. Technical assistance provided by Kristen Rayner for the scanned images. A special thanks to the science teachers who I work with, especially AMNH MAT teachers for letting me use their sketches and comments for the chapter: Chris Cubelo, Laura Carver Dionne, Meredith Fichman, Lisa Hlinka, Victoria Jones, Christina Lee, and Sean McFadden.

References

American Museum of Natural History Annual Report. (1899). *Annual report 1899* (pp. 30). http://digitallibrary.amnh.org/handle/2246/6214. Accessed.

Avraamidou, L. (2014). Developing a reform-minded science teaching identity: The role of informal science environments. *Journal of Science Teacher Education, 25*, 823–843.

Bell, P., & National Research Council. (U.S.). (2009). *Learning science in informal environments: People, places, and pursuits*. Washington, DC: National Academies Press.

Coleman, L. V. (1939). *The Museum in America: A critical study*. Washington, DC: The American Association of Museums.

Falk, J. H., & Dierking, L. D. (1992). *The museum experience*. Washington, DC: Howells House.

Griffiths, A. (2002). *Wondrous difference: Cinema, anthropology, and turn-of-the-century visual culture*. New York: Columbia University Press.

Hein, G. E., & Alexander, M. (1998). *Museums, places of learning*. Washington, DC: American Association of Museums Education Committee.

Housen, A. (1999). Eye of the beholder: Research, theory and practice. Paper presented at the conference of "Aesthetic and Art Education: a Transdisciplinary Approach," sponsored by the Calouste Gulbenkian Foundation, Service of Education September 27–29, 1999, Lisbon, Portugal.

Kisiel, J. F. (2014). Clarifying the complexities of school–museum interactions: Perspectives from two communities. *Journal of Research in Science Teaching, 51*, 342–367. https://doi.org/10.1002/tea.21129.

Leinhardt, G., & Knutson, K. (2004). *Listening in on museum conversations*. Walnut Creek: Altamira Press.

Quinn, S. C. (2006). *Windows on nature: The great habitat dioramas of the American Museum of Natural History* (p. 18). New York: Abrams.

Ramsey, G. F. (1938). *Educational Work in Museums of the United States. Development, methods and trends* (p. xiv. 289, pp. 45). H.W. New York: Wilson Co.

Sherwood, G. H. (1914, June 23). History of the relations of the American Museum of Natural History to the public schools. *American Museum of Natural History*, Miscellaneous publications, no. 5: 3. New York.

Vygotsky, L. S., & Cole, M. (1978). *Mind in society: The development of higher psychological processes*. Cambridge, MA: Harvard University Press.
Winnicott, D. W. (1965). *The maturational processes and the facilitating environment: Studies in the theory of emotional development* (pp. 37–55, Khan, M. M. R Ed.). London: Hogarth Press.
Yenawine, P. (2013). *Visual thinking strategies: Using art to deepen learning across school disciplines*. Cambridge, MA: Harvard Education Press.

Cristina A. Trowbridge is a senior manager of professional development at the American Museum of Natural History. In her role, she works with science teachers and provides support in New Teacher Induction with graduates of the Museum's Master of Arts in Teaching program. She is interested in the use of museum dioramas and exhibitions to support teacher reflection and engagement of students in science. Currently, she is researching how AMNH MAT graduates are using Museum resources and pedagogy in their instruction and affect it has on student understanding. In May 2018, Cristina received her Ph.D. in Urban Education from the Graduate Center of The City University of New York.

Chapter 12
Learning in Physical Science Opportunities at Natural History Dioramas

Sue Dale Tunnicliffe, Rebecca Gazey, and Eirini Gkouskou

12.1 Learning Everyday About Science

Children begin formal school already having experienced some everyday science in their lives. In countries such as England, many children attend preschool from 3 years of age. Increasingly, places for families offer sessions for an adult and child aged 2 years old to 'Stay and Play' (Lloyd et al. 2016) where the activities offer science learning experiences. If you observe young children at play, they are purposeful in their 'work'. They investigate, ask themselves questions and try out strategies, apparently to 'see what happens' and they try to explain what they observe in their own terms from their existing knowledge.

Psychologists assert that aspects of science are learnt in different domains; one for biology and one for physical science. Moreover, the ideas of a child about what scientists identify as physical science are different from those of scientists, hence being children's science (Hadzigeorgiou 2015). Children of different ages, and thus different stages of cognitive understanding, interpret phenomena differently. Their 'common sense' ideas are modified as they acquire new experiences and make their own observations. Additionally, they integrate information and explanations, received from a variety of sources such as teachers, into their mental model. Driver et al. (1985) suggested that the claims of young science learners are similar to those

S. D. Tunnicliffe (✉)
University College London (UCL), Institute of Education, London, UK
e-mail: s.tunnicliffe@ucl.ac.uk

R. Gazey
Museum of Science and Industry, Manchester, UK

E. Gkouskou
University of East London, London, UK

A. Scheersoi, S. D. Tunnicliffe (eds.), *Natural History Dioramas – Traditional Exhibits for Current Educational Themes*,
https://doi.org/10.1007/978-3-030-00175-9_12

of earlier societies, in particular those in western Europe. Present ideas developed in the advent of scientists who made their pronouncements based on evidence and often suffered for it, for example *Galileo Galilei* in the Renaissance in Western Europe. Whilst developmental psychologists discuss the process of learning sciences, biological science appears to be an exception to the accepted idea of how children learn; biology is different because we are members of the biological domain and are in agreement that conceptual development is a specific construction of the physical world (Inagaki and Hatano 2008). As the learner interacts with their environment they find patterns and develop a store of knowledge as past ideas are modified and developed in the light of further understanding.

Ideas are modified as fresh understanding and are assimilated. These mental models are applied regularly and some do not change even in adulthood, particularly if they have not encountered any further information. For example, a group of mothers at Sreepur Village, Bangladesh, who had not received schooling and were attending a session of 'everyday science' were adamant about their explanation of shadows and explained thus *"Sunray comes from the sun but sometime the sun swallowed the earth. Because sun is grateful to the moon. That's why earth become dark sometimes*" (translated from the Bangla by Angshuman Sarkar science facilitator). Children of different ages and thus different stages of cognitive understanding, interpret phenomena differently. Their 'common sense' ideas are modified as they acquire new experiences and observations. Thus, the same question asked to say a 6-year-old, an 8-year-old and a 10-year-old is likely to be answered differently.

Is there a place for developing physical science awareness? The ideas of early learners are not 'misconceptions', as defined by teachers well versed in the school and scientists science. This is children's science, referred to by some researchers as 'naive theories', or alternative conceptions (Driver et al. 1985). Such concepts aid the understanding of 'school science" which may lead to a 'scientists' science' (Osborne et al. 1983). It is argued that young children have the ability to acquire viable realistic concepts of the living world when involved in relevant activities where they are instructed. We argue as teachers that children, through their own observations, construct an understanding from their own experiences. Children make acute observations and explain phenomena through applying 'such personalization'.

The more similar the target for interpretation is to the human (that is themselves), the easier it is to explain biological phenomena in personal terms. Such use of the self as a template was explored extensively by researchers such as Susan Carey (1985). Hence, the use of anthropomorphic explanations or examples when visitors of all ages interpret animal structures and actions, whilst looking at animal exhibits is not surprising (Patrick and Tunnicliffe 2013). However, whilst children have some everyday knowledge of everyday animals and plants (Patrick and Tunnicliffe 2013), and know of some non-endemic animals such as lions and rhinos from media sources they have more restricted knowledge of plants (Gatt et al. 2007). Cacti, for example, were not recognised as plants, a term that is used frequently by young children synonymous with flowering plants. It follows that when looking at natural history dioramas too, visitors interpret that which is noticed in a similar way. Cognitive psychologists however, regard Scientists' science as belonging to a different domain. However, the viewers of natural history dioramas do not appear to spontaneously notice physical science in action.

12.2 Physical Science and Primary Schools

An important issue in considering physical science in particular is that in many early years settings the educators, be they carers, primary/elementary teachers or chaperones on school field trips, are not confident in their own science understanding. In particular it seems in the area of physical science and the nature of science. Indeed, in a sample of primary schools by the Wellcome Trust, only 6% of science subject leaders were found to have a degree containing any science (Wellcome Trust 2013). Furthermore, it was also stated in the UK Parliament in 2014 that only 8.3% of teachers in primary schools had science degrees (Outlined in answer to House of Commons written question 218,919, Parliament UK 2014). Thus, it is not surprising that primary teachers often feel uncomfortable teaching physical science. However, offering workshops in professional development on physical science and animals can interest teachers and provide them with more confidence to tackle simple physics, such as forces, with their classes and look for applications. Hence, looking at animals in zoos and museums is a different way of assisting children to understand some aspects of physical science in action. Furthermore, it is increasingly recognised that there are 'Big Ideas' of science knowledge and that the curriculum should be narrowed from a broad, uncoordinated content, to one of greater in-depth learning. Moreover, it is also recognised that science teaching of whatever kind should be in applied to an everyday context (Harlen 2015).

Physical science concepts are acquired in different psychological domain and hence are different to biological concepts (Inagaki and Hatano 2008). A domain is identified as a set of phenomena. So, children make sense of what they see from what they know, which can be termed 'children's science'. Some physical phenomena are given biological explanations. For instance, young children assume moving objects are endowed with biological powers such as movement at will. The possession of life is often attributed to physical phenomena such as clouds and fire. Tomkins and Tunnicliffe (2007) noticed this use of biological attribute of movement to inanimate objects in pre-secondary school aged children's responses to items on a nature table.

12.3 Recognising Physics in Action

People rely on the content of their mental models to name or identify that which they are observing. The following work described here, is a preliminary attempt to find if children can identify any manifestation of physical science in action in

Fig. 12.1 Primate diorama in Gallery 1 where these reported workshops were conducted. The collection of primates is in the diorama at the end wall. Copyright Nikhilesh Havel. (Reproduced courtesy of the Trustees of the Powell-Cotton Museum)

natural history dioramas. Of particular interest is the topic of forces, which is shown in movement or balance by the way the animals are displayed; bending their heads down to drink from a water hole for instance, or to graze on ground covering vegetation. Balance is also particularly shown in arboreal animals, such as the primates in the Primate Diorama in Gallery 1 at the Powell-Cotton Museum (Fig. 12.1). Position of eyes and ears, light and sound receptors, as well as movements and adaptations to the environment which the featured organism naturally inhabits, can also be identified.

Observing natural history dioramas spontaneously and then cued provide opportunities to identify physics in action, *albeit* at 'a moment frozen in time' as Reiss and Tunnicliffe (2011) describe in a Museum of Scotland diorama, which shows a pair of wolves frozen in their chase of a wild boar in a Caledonian pine forest.

Most animals, when alive, can make some observable movement and most possess the power of locomotion – being able to move from one place to another. Balance and centres of mass are phenomena which can be observed in natural history dioramas, as well as structures to bear the mass of the animals. Appendages, particularly legs in land-living animals, such as those of the elephant, and inquiring why the legs of these animals are much bigger in diameter than those of the antelopes or indeed of the giraffe in the diorama. Such observations can lead to the understanding of surface volume ratio and the needs associated with being warm blooded and maintaining body temperature. Observations of basic physics in action such as the optimum position for the legs, in quadrupeds then bipeds, can be developed through using modelling clay and matchsticks to try out varied positions of legs on a modelling clay 'body'.

At the Kashmir diorama in Gallery 2 (Fig. 12.2) of the Powell-Cotton Museum a boy exclaimed, *"Look, See! At that animal, he's fallen off. The wolf's made him fall off*'. A goat is depicted falling from the top of a cliff where there is a wolf as well. An observation such as this would provide an opportunity to discuss further gravity and forces.

Fig. 12.2 Kashmir Diorama in Gallery 2. Copyright Nikhilesh Havel. (Reproduced courtesy of the Trustees of the Powell-Cotton Museum)

12.4 The Galleries of the Powell-Cotton Museum and Opportunities to Develop Physical Science Understanding

The dioramas and galleries are described in detail, there is thus repetition of such in the companion book (Natural History Dioramas – Traditional Exhibits for Current Educational Themes: Socio-cultural Aspects) where a different topic is discussed but referring to the same dioramas. Gallery 1 is displaying the animals of north and west Africa, and India (shown in Fig. 12.1).

Today, Gallery 1 is the first gallery visitors see on entering the museum. However, it was actually the last gallery built by Percy Powell-Cotton himself, being completed in 1939 the year before his death. The large diorama to the left presents many species from across northern Nigeria and Chad.

The central diorama showcases the amazing diversity of Africa's primates and the different landscapes in which they live. The diorama to the back right of the gallery depicts animals from the Indian state of Madya Pradesh (which translates as 'Central Province'). The final diorama, to the right of the gallery, incorporates a

Fig. 12.3 Diorama in Gallery 3. Copyright Nikhlesh Havel. (Reproduced by courtesy of the Trustees of the Powell-Cotton Museum)

variety of landscapes and animal habitats. The far left represents the lusher woodlands around the Mkuze River, in northern KwaZulu-Natal, South Africa. The central part of the diorama, formed of a high rocky crag, represents the Ethiopian Highlands, an area where land levels rarely fall below 1500 m. The Mountain Nyala are only found in this region and have become a rare and endangered species. Finally, the desert habitat at the front of the case showcases the diversity of species found in the Sahara Desert (Powell Cotton Museum 2015a).

Gallery 2, called ‘The Pavilion’, was the first gallery designed and built by Percy Powell-Cotton and the starting point for his relationship with the taxidermist Rowland Ward, who helped build and design the museum’s famous natural history dioramas (Powell-Cotton Museum 2015b; Fig. 12.2).

Gallery 3 was the second gallery to be built, added on to the ‘Pavilion’ in 1909 (Powell-Cotton Museum 2015c). The dioramas in this gallery focus on species from equatorial Africa and the plains at the edge of these forested areas. Galleries 2 (Fig. 12.2) and 3 (Fig. 12.3) equally have the science potential to explore physical concepts but they were not used in this present pilot study.

The following table (see Table 12.1) is a summary of some of the physics concepts that are illustrated in the dioramas of the Powell-Cotton Museum. For example, gravity and forces maintaining the balance of animals and the position of the legs, especially when the neck is bent toward a water source. Linking biology with physics, for example zebra and antelope are ungulates (Artiodactyls). They are two toed hoofed animals. Their weight is distributed between two toes, and the third and fourth, which form the hooves. The one toed, or odd toed, ungulates, for example rhino (Perissodactyls), have their weight carried through by the one hoof. The various physics principles, which are manifest in the animals, are listed below. They can be identified in looking at any animals, live or taxidermy specimens, but are particularly effective to notice in a natural history diorama where the animals are exhibited in action in a realistic context (Tables 12.2 and 12.3).

Table 12.1 The Watering Hole and the Indian Forest dioramas in the Primate Gallery, Gallery 1

Diorama	Physics ideas	Animals
The watering hole (long full wall diorama on left in Fig. 12.1)		
	Camouflage	Bongo (stripes), Zebra
	Stripes	Mongoose
	Countershading	Oribi
	Pattern/sunlight	Giraffe
	Colour blending	Antelope
	Flight (forces)	Butterflies
	Reflection	Oribi
	Centre of gravity, spreading load	Giraffe
	Stray legs/heavy animal	Buffalo split hoof wider surface area sand
	Thin legs light animal	Gerenuk (on 2 legs)
	Position of legs	Quadrupeds
	Heat loss	Fur
Indian Forest (right hand side in Fig. 12.1, adjacent to the all primate end wall diorama)		
	Stripes	Tiger
	Forces (pull, pushes)	Sloth bear claws
	Forces (gravity)	Leopard & antelope in the tree
	Push	Porcupine spines (also adaption of fur/hair)
	Gravity	4 horned antelope off ground (leap up and down)

Table 12.2 Primates (Gallery 1 in Fig. 12.1, middle)

Diorama	Physics ideas	Animals
Primates		
Balancing	**Balancing:** Walking on branch	Central African red colobus, monkey
	Balancing: Sitting on the branch	Black and white colobus
	Balancing: Standing by tailing	Chimpanzee
	Communication	Gelada baboon, *Theropithecus gelada*

Table 12.3 The North Africa diorama in Gallery 1, on the right hand side nearest entrance (Fig. 12.1)

Dioramas	Physics Ideas	Animals
Gallery 1		
	Camouflage	Addax
	Camouflage (different version)	Red neck gazelle
	Balance	Addax
	Colour blending	Antelope

12.4.1 An Exemplary Guide to the Potential Physics Concepts in Gallery 1 of the Powell-Cotton Museum

Some museums have made an active effort to involve visitors with the content of their dioramas by providing artefacts related to the narrative in front of the dioramas, so visitors can not only mentally interact and interpret the narrative but also physically. The Panorama at the University of Kansas Natural History Museum, a 120-year-old, nearly 360-degree-view diorama that embodies a historic first, in the representation of nature to the public, has introduced such physical action interpretative items. This iconic exhibit represents ecological regions extending from the Arctic coast through North America into tropical rainforest. These include two interactive stations with touchable objects and activities in museum discovery guides.

Such action labels lead to the pilot studies conducted at the Powell-Cotton Museum in England to explore the spontaneous recognition of physical science in action in the dioramas and whether this could be increased by trigger workshops to refresh primary school children of science concepts they had studied.

12.4.2 Primate Diorama in Gallery 1 (Shown in Fig. 12.1)

Two pilot studies were undertaken; one with two 11-year-old boys who had studied science at their state school and were frequent visitors to this museum. The other group was of 15 mixed gender and ability 8-year-old children, half of whom had visited the museum previously. Museums in the United Kingdom run courses for parties of school children, as do zoos and botanical gardens. The majority of these courses are linked to the topics required to be studied in state schools relevant to the national curriculum of the country of residence.

The two boys were invited, each with a researcher, to look at the African diorama (on the left in Fig. 12.2) and to tell us whilst looking at the diorama, "what is it about?" Their response was of observations made with some inferences. Boy 2 replied, *"Desert. Wild variety of animals doing all kinds of things. A giraffe reaching to eat. Different species of animals. Different zebra animals, doing different things as a group"*. Boy 1 was interested in that, *"this diorama puts together animals in the same space. The diorama is like it combines different animals in arrangement for the visitor."*

His response to "Where are the things you notice?" was about the effect of the dioramas. He commented, *"It's a freeze framework of the wild, a short image, very impressive, they also have created the background representation of the wild."* Whereas Boy 2 replied, *"leaves and animals and it really seems I am there and this makes the difference (to learning science at school)"*. The responses to what the animals were doing were factual and descriptive. Interest was expressed by Boy 1 in the movement of animals portrayed but with sensitive interpretation of positions; he highlighted one antelope that was looking back over its shoulder "*as if she's lost something and she looks round to spot it. The antelope's attitude is like a tourist's attitude in a new place when confused.*"

12.5 Workshop in Middle of a Visit

In another gallery, kept for activities, the boys were introduced to the 'equipment' for a workshop, namely modeling clay and some small sticks (cocktail sticks) to represent legs. They were asked to make an animal that could stand upright and be stable with 4 legs. One boy immediately made a horizontal rectangle shape and fixed 4 legs, one at each corner of the body. The other boy decided to make a 4-legged animal with a vertical cylindrical body. This was resistant to standing up! He eventually decided to reorient his 'body' so he had a rectangular one lying horizontally. Then he fixed the legs together in the middle of the underside of the 'body'. Eventually he decided to try positioning the legs at corners and was pleased that this produced a stable model. The boys were invited to stand their 'animal' on a piece of card which acted as a 'wobble' board and to investigate how long their animal could stand for as they increased the backward and forward movements of the board. They found that by having the legs not coming down vertically from the body but at an obtuse angle (slanted) the model animals were more stable. The boys were asked to add a neck and head onto their model 'body' and then show how the giraffe-like model animal could drink by simulating the visit to the water hole, as modeled by a giraffe in the waterhole dioramas in this Gallery. They found that the animal toppled over until they had made the area between the legs wider and shallower. They remembered they had learnt about forces in school science but said it had not related to anything in their everyday world, like animal movement.

On returning to the dioramas in Gallery 1 the boys added these inherent science ideas of balance, stability and centre of gravity to their interpretation of the diorama. Boy 1 reported that "*the giraffe starts bending her legs to get her head closer to the ground. I can see stability in the animals. The legs support the position of the head. Each part of the body supports because, for example, one leg of the giraffe cannot work without the other legs.*" He noticed the information provided by the body of the animal; the spreading of legs increasing the surface area underneath, spreading their weight on their legs and nobody standing on one leg.

Boy 1 used his own body to show how an animal altered the position of its legs in order to bend down to drink and not overbalance. He was intrigued and continued identifying animals that were bending down. Boy 2 also noted that the buffalo had wide legs and the antelopes had thin stick legs. He postulated as well that animals with big ears, such as the Bongo, could hear better and needed to because it was dark in the rain forest.

The boys had also been asked to balance on one leg as a starting activity and were intrigued, particularly looking at the Gallery 1 primate diorama, to now notice that animals sitting on branches must be balanced and had to walk one leg in front of the other along a huge branch. Moreover, the study shows that children can identify science in action in animals. Thus, in addition to the usual workshops on biodiversity and conservation, basic physical science has its place in natural history museum education.

Peer group response to dioramas and the effects of a series of simple workshop activities with the 8-year-old children resulted in a greater awareness of the science in action in the dioramas. The activities tried were: to make four legged animals from modeling clay, to look at balance and stability, and to match colour and shape cards to colours and shapes of the animals (see Table 12.4).

The responses of 3 groups of 8-year-old primary pupils to the colour matching activity and a shape matching activity are presented at the Tables 12.4 and 12.5 below.

Also, laminated cards of the outline of simple geometric shapes enabled the children to find the basic shape in the animals and name the animals (see Table 12.5).

Finally, the museum educator had constructed a long ‘line’ out of thick paper an inch wide, which was adhered to the floor. Children were asked to walk along it

Table 12.4 The response of boys, girls and mixed groups to the Colour activity, carried out in gallery 1

Colour	Animals		
	Boys group	Girls group	Mixed group
Pink	Giraffe tongue	Gelada Baboon, Gorilla’s mouth.	
Red	Baboon	Gorilla’s eyes	
Beige	Donkey, Addax, Dorcas Gazelle	Western Lowland Gorilla	Gazelle, Ass, Oryx.
Brown	Swayne’s Hartebeest, Roan Antelope, Beetle	Porcupine, Mona Monkey, Bush Pig, Addax, Talapoin, Guenon, Zebra, Gorilla’s teeth.	Suni, Lechwe, Wildebeest, Giraffe.
White	Black and white Colobus, stripes on the Nyala.	Colobus Monkey.	
Black	Sloth bear, Bush pig, Cape Buffalo, Monkey.	Bear, Gazelle’s horns, Udad’s horns, Colobus Monkey, Chimpanzee, Gorilla, Bush Pig, Cape Buffalo	Gorilla, Buffalo, Chimpanzee, Colobus, Bush Pig, Giant Forest Hog.
Orange	Crab	Tiger, Leopard	
Black and white stripes	Zebra, Beetle, black and white Colobus, Tiger’s face	Zebra	Zebra, Colobus
Cream with brown splodges	Giraffe, Cheetah	Giraffe	

Table 12.5 The responses of boys, girls and mixed groups to the Shape activity carried out in Gallery 1

Shape	Animals		
	Boys group	Girls group	Mixed group
Circle	Eyes, pond, leaves, Mrs. Gray's Lechwe	Eyes, Sloth Bear's head, Hyena's nose, Buffalo's nostrils.	Monkey's faces, Monkey's noses, Tiger's feet, Nest.
Semi circle	Zebras, ears, bananas.	Gazelle's nose, Tiger's ears, Buffalo's legs.	Hooves, bananas, claws, Monkey's nose, Ass' nostrils, Animal's nostrils.
Oval	Leaves, Deer's body, eggs, Mongalla Gazelle.	Buffalo's ears.	Boar's nose, leaf, noses, bodies, ears, faces, eyes, Beetle, Tiger's tail.
Cylinder	Giraffe neck, tree branch, nest, legs, horns, Monkey torsos.	Buffalo's legs.	Animal's legs, feet, tree, horns, tails, paws, claws, bananas.
Rectangle	Monkey torso, head, rocks.	Edmi Gazelle's body, Sloth Bears claws, Baboon's foot, Bush Pig's leg.	Rocks, bodies, Ass' body.

normally, finding they had to put one foot in front of the other to stay on the line, and then on tip toe and keep their balance. Children found that when walking on tiptoe they needed to use their arms in order to maintain their balance.

The lesson that emerged for us from these preliminary workshops is that primary science is not taught within a meaningful context in primary schools. The children needed a preliminary view then a workshop or in situ activities and a second viewing of the same diorama when the children are able to identify science in action.

12.6 Conclusion

Physical science principles are implicit in observing the living world and these life-sized representations of a moment in time, whether a faithful representation of a known scene or a conceptual construction diorama illustrating biogeographic principles provide the opportunity. School science, in the primary school at least, is not taught within a familiar context for children and they do not use school-learnt knowledge when interpreting, in this case natural history dioramas, until they have been cued into the science concept with some 'hand-on' activities. The plants and animals included in natural history dioramas display physical science in action through the positions in which they are exhibited; upright, flying, swimming, camouflaged, bending or balancing for example, as well as providing clues for visitors to recognise the earth science element from the habitat portrayed. Biodiversity learning can be amplified by the recognition of manifestations of other sciences within any one diorama.

References

Carey, S. (1985). *Conceptual change in childhood*. Cambridge, MA: Bradford Books, MIT Press.

Driver, R., Giene, E., & Tiberghien, A. (1985). *Children's ideas in science*. Buckingham: Open University Press.

Gatt, S., Tunnicliffe, S. D., Borg, K., & Lautier, K. (2007). Young Maltese children's ideas about plants. *Journal of Biological Education, 41*, 117–121.

Hadzigeorgiou, Y. (2015). Young children's ideas about physical science concepts. In K. C. Trundle & M. Sackes (Eds.), *Research in early childhood science* (Vol. 67). Dordrecht: Springer.

Harlen, W. (2015). *Working with big ideas of science education*. Science Education programmed (SEP) of IAP. intteracademies.net/activities/projects/12250.aspx. Accessed 20 Feb 2016.

Inagaki, K., & Hatano, G. (2008). Conceptual change in naive biology. In S. Vosniadou (Ed.), *International handbook of research on conceptual change* (pp. 240–262). New York: Routledge.

Lloyd, E., Edmonds, C., Downs, C., Crutchley, R., & Paffard, F. (2016). Talking everyday science to very young children: A study involving parents and practitioners in an early childhood centre. *Early Child Develoment and Care, 187*(2), 244. https://doi.org/10.1080/03004430.2016.1226355.

Osborne, R. J., Bell, B. F., & Gilbert, J. K. (1983). Science teaching and children's views of the World. *European Journal of Science Education, 5*, 1–5.

Parliament UK. (2014). *Answer to house of commons written question 218919*. Retrieved from http://www.parliament.uk/business/publications/written-questions-answers-statements/written-question/Commons/2014-12-16/218919/. Last Accessed 15 Dec 2017.

Patrick, G. P., & Tunnicliffe, S. D. (2013). *Zoo talk*. New York: Springer.

Powell-Cotton Museum. (2015a). *Gallery 1*. Retrieved from http://www.quexpark.co.uk/museum/museum-galleries/gallery-1.html. Last Accessed 15 Dec 2017.

Powell-Cotton Museum. (2015b). *Gallery 2*. Retrieved from http://www.quexpark.co.uk/museum/museum-galleries/gallery-2.html. Last Accessed 15 Dec 2017.

Powell-Cotton Museum. (2015c). *Gallery 3*. Retrieved from http://www.quexpark.co.uk/museum/museum-galleries/gallery-3.html. Last Accessed 15 Dec 2017.

Reiss, M., & Tunnicliffe, S. D. (2011). Dioramas as depictions of reality and opportunities for learning in biology. *Curator the Museum Journal, 54*(4), 447–459.

Tomkins, S., & Tunnicliffe, S. D. (2007). Nature tables: Stimulating children's interest in natural objects. *Journal of Biological Education, 41*(4), 150–155.

Wellcome Trust. (2013). *The deployment of science and maths leaders in primary schools. A study for the Wellcome Trust.*

Sue Dale Tunnicliffe is a Reader in Science Education at UCL Institute of Education, University College London. She holds a bachelors degree from Westfield College, University of London in Zoology and trained as a secondary biology teacher at the university's Institute of Education. After teaching in grammar schools for a few years she had children and, wanting to find out more about how young children learnt science, particularly biology, she entered primary school teaching and is now interested in pre school learning, as well as animals as exhibits. She set up and ran a new Primary Science and Design Technology advisory team for London Borough of Richmond and then became Head of Education at the Zoological Society of London. She has worked at the BBC and in a cultural museum. Her doctorate was from King's College, London'Talking about animals: conversations of children in zoos, a museum and a farm". She has published widely.

Rebecca Gazey was the Head of Learning at Powell-Cotton Museum, House and Gardens, Kent (UK). Rebecca studied American Studies (with International Study) at the University of Nottingham. Upon graduating she began volunteering in the heritage sector and was employed by the Royal Air Force Museum, Cosford as a Learning Assistant. Whilst at the RAF Museum she undertook a Masters in Museum Studies at the University of Leicester. Upon the completion of her studies, Rebecca joined the Royal Engineers Museum, Library and Archive in Gillingham as their Community Engagement and Learning Officer and had joined Powell-Cotton Museum, House and Gardens. Rebecca's specialism within her career, academically and professionally, has been museum education with particular focus on making heritage accessible to children under 5.

Eirini Gkouskou is a Senior Lecturer in Early Childhood & Education Studies at Cass School of Education and Communities at University of East London, a Research Associate in Science Education at UCL Institute of Education, University College London and a Deputy Editor of the Journal of Education in Museums. Formerly, a Researcher at the University of Patras, Greece, and senior teacher of Preschool Education at settings both at United Kingdom and Greece. Her research interests concern epistemological and educational aspects of the transformation of scientific knowledge to school science in formal and non-formal educational settings, Inquiry Based Learning and Didactic Transposition of the learning outcome of the Natural History Dioramas.

Chapter 13
Learning Science Through Encounters with Museum Dioramas
Themes and Patterns in Students' Conversations

Jesús Piqueras, Karim Hamza, and Susanna Edvall

13.1 Introduction

For many years, museums have been recognized as environments that encourage informal learning experiences of individuals, families and school children. In many of these venues, particularly in natural history museums and science museums, the natural history dioramas, three-dimensional depictions of animal-landscape sceneries (Kamcke and Hutterer 2015), are still one of the most essential components in the exhibitions. During the last years there has been an increasing interest of the study of dioramas as a resource for learning and teaching, however the number of studies focusing on the learning process, where visitors or school students interact with such dioramas is limited (Ash 2004; Tunnicliffe 2015; Tunnicliffe and Scheersoi 2015). In out-of-school museum experiences, dioramas can be used as an arena for observation, exploration and discussion of different natural phenomena (Tunnicliffe 2015).

As a realistic representation of animals and plants in their environment, dioramas provide the learners with an opportunity to observe, ask questions and seek more information (Ash 2004). In such opportunities, students can establish relationships between their own experiences and the artefacts of the diorama and their articulation of tentative explanations for the displayed scene (Piqueras et al. 2008). Here, we present a study of science learning, on a discursive level, in a teaching activity for a museum of natural history. The teaching activity utilizes a variety of dioramas with preserved animals in scenes that reproduce their natural environments and

J. Piqueras (✉) · K. Hamza
Stockholm University, Stockholm, Sweden
e-mail: jesus.piqueras@mnd.su.se

S. Edvall
Swedish Museum of Natural History, Stockholm, Sweden

A. Scheersoi, S. D. Tunnicliffe (eds.), *Natural History Dioramas – Traditional Exhibits for Current Educational Themes*,
https://doi.org/10.1007/978-3-030-00175-9_13

behaviours. The theoretical framework that we used in this study is the practical epistemology analysis (Wickman and Östman 2002; Wickman 2004; Kelly et al. 2012). In a previous study (Piqueras et al. 2008), we introduced this method as a high-resolution approach of studying a group of student teachers' moment-to-moment learning in the activity with the dioramas. We used then only a minor part of the conversation of one group of students. In a later study (Piqueras et al. 2012) we provided an account of the entire activity with analysis of the whole conversation in the same group, illustrating how practical epistemology analysis could be used to also examine the development of discursive themes in students' reasoning. Here, we have scaled-up our analysis to ten different groups of student teachers' conversations. Through comparing several groups of students in this study, we have not only explored and described the diversity and frequency of the identified themes, temporal patterns and relationships between themes in conversations, but also discussed the role of the teacher and use of questioning against other studies reporting on learning in different informal environments.

13.2 Methods

13.2.1 The Teaching Activity

The data presented in this study comes from an activity of the educational program for teachers at the Swedish Museum of Natural History in Stockholm. In this program, teachers have the opportunity to learn more about informal settings and explore the use of different exhibits in the museum. One important aim of the program is that these teachers can then perform similar activities with their own students during a school visit to the museum. The teaching activity chosen for this study utilizes various dioramas of the permanent exhibition Swedish Nature. These dioramas use preserved animals (taxidermy mounts) and reproduce scenes of their natural environments and behaviours. At the beginning of the activity, the participants are assigned to small groups and asked to study one of the dioramas of the exhibition. During this 5 min phase, the participants are encouraged to write down some questions elicited by the study of the diorama. The groups are then invited to discuss the questions during a 10–15 min period. During this activity, the museum educator or the school teacher will join the group discussions to support participants through short interventions. In the final part of the activity, each group presents an account of their observations and questions to the rest of the class. During this presentation, the questions elicited by the diorama are re-examined and discussed with the museum educator and the teacher.

The participants in this study were student teachers (hereafter abbreviated as "students") at the end of their university studies who will teach multiple subjects in the Primary School. Typically, four to five dioramas are used in the activity, but for this study we videotaped ten different groups working in the same diorama on different occasions. Eight out of the ten groups have three students and two groups have four students. In the study, one of the authors was the teacher interacting with the group. Apart from the video recordings, conversations were audio recorded with individual voice recorders. Students' talk was transcribed verbatim and translated from Swedish into English with as small changes as possible from the original wording.

13.2.2 The Diorama

The diorama used in this study shows a female white-tailed eagle (*Haliaeetus albicilla*) eating the remains of a roe deer in a snowy landscape (Fig. 13.1).

The display diorama has a hexagonal shape and allows the study of the objects and details inside from different perspectives. In addition to the white-tailed eagle, there are a pair of hooded crows (*Corvus corone cornix*), one of them staying at a prudent distance from the eagle, whereas the other pinches the eagle's tail-feathers (Fig. 13.2).

The diorama is inspired by a real scene that is shown in a short video sequence on a monitor screen close to the diorama, though, this resource was kept hidden from the student teachers during the activity. The main intention of the diorama is

Fig. 13.1 Hexagonal diorama showing a white-tailed eagle eating the remains of a roe deer and pair of hooded crows, one of them staying at a prudent distance from the eagle, whereas the other pinches the eagle's tail-feathers

Fig. 13.2 Detail of the diorama showing the crow irritating the white-tailed eagle

to show the co-operating behaviour of the crows to steal food from the eagle (Lars Bern, curator; interview, December 16, 2009). Another explicit curatorial purpose of the diorama is to challenge the curiosity and imagination of the visitors by placing additional details and clues in the main scene. Thus, there are yellow spots and footprints interspaced on the artificial snow (Fig. 13.3) belonging to a presumptive fox which has visited the carrion before the eagle and the crows, and the head of the roe deer is missing (a common behaviour of foxes is to take away the head of the animals they predate).

The landscape represented in the scene is inspired in a rock formation, typical for the maritime archipelago at Stockholm, where these eagles are relatively common (Lars Bern, curator; interview, December 16, 2009).

13.2.3 Analysis of the Conversations

We analyzed the students' conversations using the practical epistemology analysis developed by Wickman and Östman (2002; Wickman 2004). Four analytical concepts are central in practical epistemology analysis: *encounter*, *gap*, *relation*, and *stand fast*. Encounters occur between persons and between persons and artifacts or natural phenomena. As an activity proceeds, the participants notice gaps as a result of such encounters. To fill a gap, participants establish relations to those things which they are already familiar with, and which they do not need to put into question for the moment. These are the things which *stand fast* in the particular situation. Usually, that which stands fast is seen through those words which are used by the

Fig. 13.3 Detail of the diorama showing the carrion, the footprints and the yellow urine spots on the snow from a fox that visited the carrion of the roe deer

interlocutors without questioning. An example of how these concepts are used in the analysis is provided in the following excerpt where Marta, one of the students in group 2, noticed a gap about the carrion in the *encounter* with the diorama.

Marta: I don't think it's a roe… what's the name… a roe deer he [the eagle] has killed. It looks more like a badger or something like that.

Josephine: But roe deer have also such white tufts there behind. Besides, I think that badgers are more black and white.

The first *gap* concerns the identity of the carrion (what kind of animal it is) and Marta establishes a *relation* to fill it ("it looks like a badger"). At the same time that Josephine notices another gap in the encounter with the diorama (the tuft), she establishes a relation to fill the gap, grounded in her experiences of the features of roe deer and badgers (the white tuft-the roe deer and badgers-black and white). Several words *stand fast* to Marta and Josephine, that is, words that they use without questioning. These words, for example: roe deer; badger; and tufts are used as starting points to establish *relations* to fill the *gaps*.

It is important to point out that these four concepts of the practical epistemology analysis are strictly analytical and operationalized in relation to each other, instead of being related to anything particular to the participants. For instance, in our example, it is irrelevant to ask whether Marta "really" noticed the *gap* or not in the *encounter* with the carrion. If Marta established at least the *relation* ("it-looks-badger") then by definition, this means that the gap has been noticed. The same is valid for the concept of *stand fast*. It is irrelevant to ask whether the words like roe deer, badger or tufts "really" stood fast to Marta and Josephine, because it is not a claim about what they thought or understood. It only implies that these words (or actions) observed in the students' discourses are not – for the moment – questioned by them. In this way, the practical epistemology analysis rescinds the need to make inferences about what is "actually" going on in a situation, or what the participants "really" mean.

The transcriptions of the students' conversations were coded by marking stand fast, encounters, gaps, and relations and divided into a series of steps, where the criteria for segmentation were the themes discussed during the conversation. Hence, a new step was initiated when the students began a new theme or switched from one theme to another. In the continuation of the previous excerpt we exemplify how the conversation in this group switched into a new theme, the footprints of the snow.

Marta:	Yes, but you can't see any legs [chuckling]
Josephine:	Of course! They have been "cut down" [chuckling]
Marta:	And the head is missing too.
Mikaela:	Aha.
Marta:	But here you can see footprints from an animal and they are neither from the bird. It isn't a roe deer or a predator that has walked here. They look more like dog paws.
Josephine:	Yes, you are right.
Mikaela:	And they are from them [points the crows]
Marta:	No, that's a crow. And they aren't from the eagle, because the eagle has claws. Neither from a roe deer, because they have hooves.

In the first turns of the excerpt, Marta and Josephine continue to talk about the carrion and noticed two new gaps (the head and legs of the carrion are missing) and the gap concerning the identity of the carrion lingers in the conversation. To proceed with the activity, Marta noticed a new gap, the footprints on the snow and then they try to establish relations to fill this gap ("footprints-not from the bird", "footprints-dog paws", "eagle-has-claws", "roe deer-have-hoof"). Thus, we identified in this part of the conversation two different steps, the first corresponds to the theme the carrion and the second corresponds to the theme of footprints of the snow (group 2, steps six and seven, see Fig. 13.4).

Thus, the analysis of the conversation using practical epistemology analysis and the subsequent thematization allows us to study which encounter with the diorama were difficult for the students to make meaning of. For instance, in the previous example, we can conclude that the gap corresponding to the identity of the carrion remained unfilled and lingered in the conversation, whereas the second gap, the footprints on the snow, was filled with relations resulting in a plausible explanation for the students.

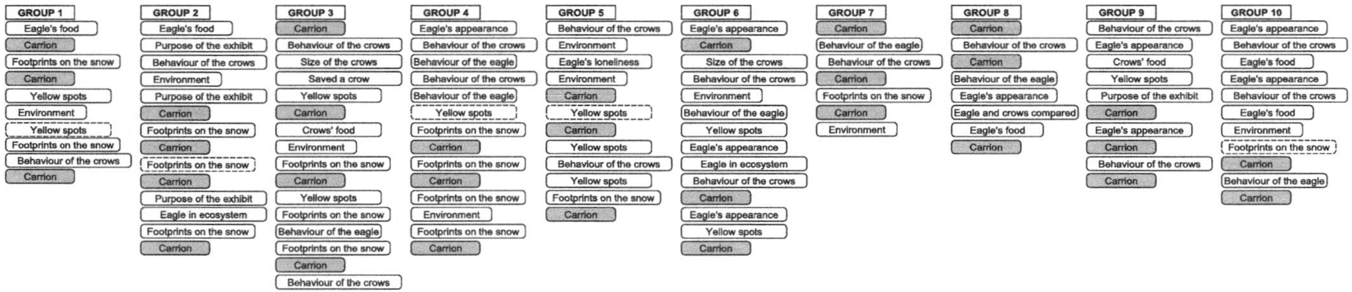

Fig. 13.4 Overview of the student teacher's conversations. The conversation of each group is represented as a chronological sequence of steps, where different themes were treated, i.e. in group 1 the conversation started with the theme Eagle's food and ended with the theme Carrion (shaded box). Themes drawn in discontinuous line denotes teacher's intervention

13.3 Findings

The ten students' conversations elicited during the educational activity with the diorama were, on average, 12 min long (range 09:40–13:48). The length of the conversations measured as a number of steps varied from seven, for group 7, to sixteen for group 3 (Fig. 13.4). All over, 15 different themes appeared in the conversations, showing groups 2 and 4 the highest number (9) and group 7 the lowest (5). The number of different themes per group was positively correlated with the length of the conversation measured as steps ($P < 0.001$, Spearman's Test) but neither the number of steps ($P = 0.758$) nor the numbers of themes ($P = 0.973$) were correlated with the duration of the conversation in minutes.

The occurrence of the 15 different themes across the 116 steps in the ten conversations showed a skewed distribution pattern (Fig. 13.5), where the most common theme treated by the students was related to the carrion of the roe deer (31). This theme was twice as common as the second most frequent themes, the behaviour of the crows (16) and footprints on the snow (15). Relatively frequent themes were the yellow spots on the snow (11), the environment represented in the exhibit (9) and the eagle's appearance (9). Less common in the conversations were the themes behaviour of the eagle (7), the eagle's food (5), and the purpose of the exhibit (4). The six remaining themes occurred only on one or two occasions (Fig. 13.5).

Even though each conversation developed in different ways, it was possible to identify certain patterns in the order that themes appeared. Five groups started the conversation noticing gaps related to the eagle appearance or the eagle's food:

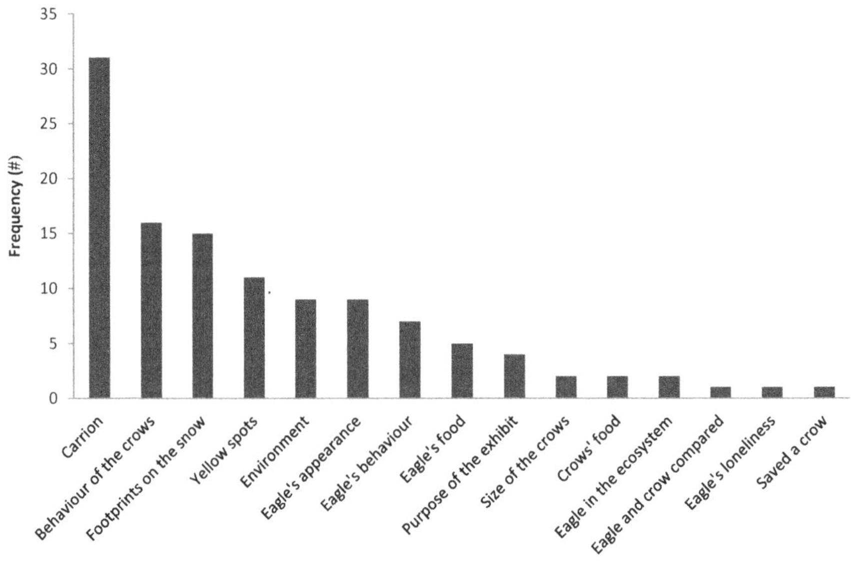

Fig. 13.5 Absolute frequency of the 15 different themes in ten student's conversations

Group 1
Charlotte: Firstly… what he is eating. He is a flesh-eater [the eagle].

Group 2
Marta: The first I thought… it was… it was dead. [the carrion].

Group 4
Samuel: I think that it looks like a very big animal eating everything…

Group 6
Hanna: He is really impressive! [the eagle].

Group 10
Lisette: The first thing that struck me was that it is *so large* [the eagle].

Even though these themes addressed two very visible aspects of the exhibit, they were not often revisited later in the rest of the conversation (Fig. 13.4). However, five groups started by noticing specifically the identity of the carrion or the behaviour of the crows:

Group 3
Lena: But, what I thought, it is… well… the eagle is feeding on a roe deer.

Group 7
Tindra: What do you think about?
Jonas: I figured out what kind of prey it is.
Anna: Ah, I think it is a fox or a lynx, may be.

Group 8
Anna: The first question…
Johanna: Mm.
Anna: What kind of animal he is feeding on?

Group 5
Tuva: The crow… that he dares to pinch… at the tail's feathers.

Group 9
Ebba: What crossed your mind to begin with?
Helena: I looked at… the crow there… it is pulling the tail's feathers.

The next most frequent themes, footprints on the snow (hereafter abbreviated as "footprints") and yellow spots on the snow (hereafter abbreviated as "yellow spots"), related to more conspicuous details in the exhibit and appeared later in the conversation (Fig. 13.4). Only one group (1) noticed the footprints earlier in the steps of conversation that the fourth step, and no group noticed the presence of the yellow spots until the fourth step (Fig. 13.4). Occasionally, these themes were introduced by the teacher (Fig. 13.4) to help the conversation go further, but on no occasion did the teacher provide further explanations. We present here some excerpts that show the first turns of these themes:

Group 1 (step 3, footprints)
Charlotte: Here there are footprints. What kind of footprints?
Isak: Are there footprints? No idea!

Group 2 (step 7, footprints)
Mikaela: But here you can see some footprints, did you hear me? From some animal and it is not a bird… anyway, not from a roe deer walking around here. They are more dogs paw-like.

Group 3 (step 5, yellow spots)
Cecilia: I wonder what is that yellow pouring there.
Lena: Yes, exactly…
Djamila: It can't be blood.
Cecilia: It looks like somebody has done a wee-wee.

Group 4 (step 6, yellow spots)
Camilla: Then, there is the question… it is… whose is the urine? Or is the yellow something else?

Group 5 (step 6, yellow spots)
Teacher: Have you noticed the snow?
Tuva: Yes, we have written about that… we wrote: what is the yellow on the snow?

Group 6 (step 7, yellow spots)
Sara: Then I thought that that must be urine... or something else?
Hanna: Yes… I guess so.
Cecilia: Yes… or it's somebody's gall bladder that burst… I don't know.

Group 7 (step 4, footprints)
Jonas: I see other footprints on the snow from another animal.
Tindra: Aha, they are small paw prints.

Group 9 (step 4, yellow spots)
Frida: Then, I looked a bit at the yellow… I wonder what it is.

Group 10 (step 8, footprints)
Janette: But here… That's great! Imprints on the snow, I haven't found them until now.

These patterns of conversation suggest that there are certain parts in the diorama that act as triggers for the conversation (the eagle's appearance, the eagle's food, the behaviour of the crows and the carrion), they act as starters to initiate the first observation and conduct the initial reasoning in the group. Contrastingly, the second group of themes appear later in the conversation (footprints and yellow spots) and correspond to less visible or evident parts that need careful and focused observations of details in the diorama. These two themes, however, hold a crucial role in the students' reasoning when they tried to make meaning of the most challenging detail of the diorama, the carrion.

The carrion eaten by the eagle in the diorama was the most difficult part for all groups of students during the activity. The carcass of an animal consisting of fur, several bones and a missing skull was noticed spontaneously by all groups, which was consistent with our previous study (Piqueras et al. 2012), the immediate gap was about the identity of the carrion. A more detailed analysis of the conversations showed that this theme was initiated in six out of the ten groups, by relations to the theme the eagle's food or the theme the eagle's appearance.

Group 1

Charlotte: First of all … he's a carnivore [the eagle].

Isak: Is he a carnivore?

Charlotte: Yes.

Isak: I was wondering about what he's eating … is it dead, or isn't he who killed it [the carrion]?

Group 6

Hanna: It's stunning! [the eagle].

Gunnel: Mm.

Hanna: Firstly, I thought, such a power. It's a rather large animal though [the carrion].

Sara: What's sort of animal? I can't see what it's.

Hanna: No, you can't, can you? But you can still see that it's…

Gunnel: You know, I thought first that it was a rabbit, but when you look here [the footprints]…

Sara: The footprints, yes.

Gunnel: Yes, the footprints… it could be some fox, or some…

Hanna: Yes, a wolf may be… it's something big though… such a power a bird has!

The excerpt of group 6 provides a good example of the reasoning articulated in several groups. They started talking about the food or the appearance of the eagle and noticed the gap of the carrion. To fill the gap they try to establish relations to previous experiences and noticed new gaps, as in this case the footprints. The students filled the gap of the footprints ("it could be some fox", "a wolf maybe") but it is clear that they could not make meaning of the identity of the carrion, thus the gap remained.

Another example of how the themes footprints on the snow and yellow spots became intertwined in the reasoning, was provided by the conversation of group 4. Together with the footprints, the theme yellow spots was also present in the students' conversation and helped them establish relations to the identity of the carrion. The students firstly noticed the yellow spots (seventh step in the conversation, Fig. 13.4) and this led the conversation to the theme the footprints.

Ann: I thought that there isn't so much blood around… the animal [the carrion]. Therefore, I reckoned that he [the eagle] didn't bring it down… Mm. Then, there is this yellow on the ground, it can be wee-wee.

Camilla: I wrote it down too… urine… whose is it? Maybe it's something else, however, since there are other marks [footprints] it could be another animal that marked the carrion, I reckoned.

The group continues reasoning about the footprints (ninth step in the conversation, Fig. 13.4), establishing relations to fill the gap related with the animal that have left them.

Camilla: There are marks of paws around, it could be a carrion-eater too. A fox or something else.
Samuel: A wolf or a fox.
Camilla: A wolf? These steps are too small, too small for a wolf tracks.
Ann: I believe so. It looks like a dog, dog-size. You know… not a dog, but similar.

The students continued reasoning and noticed the carrion again (step 10, Fig. 13.4) and used the relationship they had established in the previous steps with the footprints on the snow ("A wolf or a fox"). Initially Samuel suggests that the carrion could be a fox, but noticing again the size of the footprints and observations of the anatomy of the carrion allowed the students to fill the gap establishing new relations.

Camilla: What kind of animal is it [the carrion]?
Samuel: Maybe it is the fox that died, it looks like skin from a fox. The fur looks like it, doesn't it?
Ann: Mm.

Since the students could not go further with their reasoning, the teacher intervened suggesting a closer study of the back part of the carrion.

Teacher: If you look at the back part [the carrion]…?
Camilla: Well, I don't know… but it [looks like] roe deer.
Samuel: Aha, that's it.
Ann: Aha, it's a white spot there, a white spot on the tail.
Camilla: A fox doesn't have such a long neck… or neck bones…
Samuel: No, it doesn't.
Camilla: So, maybe it's a roe deer.
Samuel: Yes, surely, I believe it's a roe deer. It has a white spot just there.
Camilla: Mm… and I thought the neck… too long neck for a fox…
Samuel: No, certainly not. And those aren't footprints from roe deer, for sure.
Beatrice: No.
Samuel: They have claws too [the footprints]
Ann: Yes.
Samuel: I believe it's a wolf.
Beatrice: A wolf?
Camilla: I think that they are too small.
Ann: Yes, I think so, too.

Interestingly, the theme the footprints is revisited by the students in the 11th step (Fig. 13.4) after they have filled the gap of the identity of the carrion. They are not completely sure about the animal that left the footprints ("A wolf or a fox"), but by establishing a relation to the carrion, they can fill the gap that there is another animal in the scene represented in the diorama ("those aren't footprints from roe deer, for sure"). In similar ways, establishing relations to the footprints and/or to the yellow spot to fill the gap of the identity of the carrion was found in all groups. Only in one group (3), the gap related to the identity of the carrion is filled immediately in the beginning of the conversation since one of the participants, Jenny, recognized that the carrion belongs to a roe deer.

Jenny: Ok, does it awake any associations?
Lena: I was thinking about, well… it's the eagle, sitting and eating there on a roe deer.
Jenny: Mm.
Djamila: Mm.

Identifying the carrion as roe deer early on in the activity, seems to have important consequences for the rest of the conversation in this group. They spend more time visiting other themes and the carrion helped them to establish new relations that provide a consistent interpretation of the scene. One example is when they discuss how the carrion of the roe deer has end up in the scene.

Lena: But, is it [the roe deer] a natural prey for a white-tailed eagle? I don't know… may be.
Jenny: I believe that they catch smaller animals…
Lena: I reckon they can take small roe deer too, if they got stuck in the snow or something like that.
Lena: Someone has put it out for sure. Yes, I believe so.

In our previous study of group 5 (Piqueras et al. 2012), we found that anatomical observations of the carrion played an important role in the efforts of the students when they tried to fill the gap of the carrion's identity. When analysing the ten groups' conversation in this study, we found that some relations to anatomical parts of the carrion were present in all groups. Noticeably, the long neck of exposed bones of the carrion discussed by the students in the previous excerpt ("I thought the neck… too long neck for a fox", Camilla) was explicitly noticed by eight out of the ten groups. Furthermore, in two of the conversations (groups 6 and 8) this relation was crucial in excluding the possibility that the carrion was a fox or another animal, even though other relations associated to anatomical details were important, i.e. the colour of the fur and size of the carrion.

Group 6
Sara: It has a rather long neck, hasn't it? ... But what sort animal does have such a long neck?

Group 9
Helena: Yes… a roe deer, you see the neck there.
Charlotte: It might be… and the chest.

Group 6
Hanna: If you think about, the fur… it's rather like…
Gunnel: Yes, I thought about a fox tail though, but this [the carrion] is white on the belly.

Group 10
Lisette: It could be a lynx too. We look at it.
Emely: But, it's a little bit reddish, isn't it?

Group 7
Jonas: You know, I thought it was a hare first, but when I saw the skeleton there, it feels that it's something larger.

The behaviour of the crows was a theme present in all ten groups' conversation (Fig. 13.4). All groups spontaneously noticed the presence of the crows in the diorama and established analogous relations to explain the implied behaviour of the crows. Words such as "irritating", "bullying", "mocking", "annoying" and "pinching" were used in these relations to fill the gap describing the actions of crows against the white-tailed eagle. Furthermore, in all groups, noticing the irritating behaviour of the crows was an important observation to establish relations between the competition for food (the carrion), the crows and the eagle displayed in the diorama.

Group 1
Ebba: I thought… that little bird goes for the big one.
Anja: Now I understand what the crow is doing there. He wants access to the carrion.

Group 2
Josephine: Aha… what do that birds [the crows] want? You see, they want a bit too.
Marta: And the crow tries to pinch the eagle at the tail's feathers.
Josephine: To get him away.

In group 5 the students established similar relations to fill the gap in the first step of the conversation, that is, the crows irritate the eagle to get a bit of the carrion, but when they revisited the theme in the ninth step, some minutes later, they found alternative explanations for the behaviour of the crows (Piqueras et al. 2012).

Elisabet: It is really bold as you said.
Tuva: Yes. It was the first I noticed…God that he dares! [the crow].
Eva: I thought…they live in symbiosis.
Tuva: Exactly.
Eva: [laugh]
Tuva: What do you mean? ...that he's eating?

Eva: He [the crow] helps him [the eagle] pulling out the feathers that are loose…

Tuva: Aha!

Eva: …and then take the rest, but I believe more in your theory…that he's trying to scare him away.

Tuva: That he is trying to drive him away…yes…but, don't be sure about that… It may be that he's keeping him tidy.

Eva: Mm.

Thus, the irritating behaviour of the crows is not only explained in terms of competition for food, but also as a symbiotic relationship between the eagle and the crows. Here we see how a gap can be noticed on different occasions in the course of the conversation. Thus, gaps are neither limited to determined relations nor to a linear development of the reasoning. Rather, in the encounter with the diorama the students' reasoning develops dynamically, old gaps being revisited and new relations being established.

13.4 Discussion

The results of this study clearly show that the combination of the diorama as resource for teaching and the design of an activity based in the students' own question offers important learning opportunities. Analysing students' conversations in this way makes it possible to view their learning as a journey where the different themes represent alternatives paths for the direction of the learning process (Piqueras et al. 2012). In this process the content raised in the different themes in form of biological terms, concepts and reasoning is essential to describe the student's meaning making in the encounter with the diorama. Even though the purpose of the activity can condition the content of the conversations between these students, there exists vast evidence from visitor studies, particularly families, showing that these conversations are extraordinary rich in biological content when such groups of visitors interact with dioramas and other biological exhibits (Allen 2002; Ash 2003; Palmquist and Crowley 2007; Tare et al. 2011; Tunnicliffe 2008; Zimmerman et al. 2010).

Although limited to the analysis of the conversations of ten groups, the results of this study provide some interesting insights of the potential of dioramas as a resource in teaching activities. We found that the number of themes elicited by the students' own questions in the diorama encounter was limited, and relatively few themes dominated the students' conversation during the activity. However, even though some groups spent more time than others in the conversation, our analysis showed that the richness of the conversation, in relation to the number of different themes discussed, was not determined by the time spent in the conversation. Such conversational richness suggested that some groups developed a more active exploration of the diorama. Also, students' prior experiences and knowledge - as occurred in group 3 with the immediate recognition of the roe deer - might increase the possibilities to explore other details of the diorama. Understandably, it is difficult to say whether the absolute number of 15 themes reflects all the diversity of potential conversations for this particular diorama. However, our results confirm the observations and experiences of the museum educators over many years, namely that the observations, questions and themes elicited in the activity, are rather limited and predictable, even though new questions can arise from time to time. Similarly, other studies have shown that although patterns of museum dialogue may seem unique to each family, they are not entirely idiosyncratic (Ash 2003). However, it should be noted that the participants in this study come from a rather homogeneous group of people, namely young student teachers, participating in an educational activity with particular purposes.

The most frequent *gap* noticed in the encounter with the diorama concerned the identity of carrion. This result is remarkable considering that the main purpose of the diorama, the curatorial intent, was initially to show the cooperation of crows. Although this outcome confirms the results suggested in our previous study where one conversation was analyzed in more detail. A previous study (Scheersoi 2015) reported that particular details in dioramas are more likely to evoke emotional responses that capture and hold visitors' attention. Our study findings suggest that what captured the students' attention was the type of animal the carrion was, which resulted in numerous lingering gaps and developed as a central theme in all groups.

In contrast, the second most frequent theme, the behaviour of the crows, was easy to make sense of by the students. That is, all the groups were able to provide an interpretation of the scene represented in the diorama in terms of ecological relationships between the eagle and the crows. Where previous studies of conversation of families in museums and aquaria have shown that the family interweaves different thematic areas (Ash 2003; Ash et al. 2007) but certain issues, as in our study, appear to have a pivotal role in learning interactions, for example, differences between living and non-living elements in the exhibition (Ash et al. 2007). Similarly, in a study of school students' conversations in a museum visit, Gilbert and Priest (1997) found that certain identification of particular relationships (incidents) were critical for the continuation of a discourse either because they enabled greater meaning to be attributed to past experience or because they acted as a simple bridge to later activity.

When analysing the conversation, the *gaps* noticed in the encounter with the diorama, and the *relations* students made to fill them, allowed us to identify some noticeable patterns in the student's reasoning. The conversations started with different themes, some of them were more frequent and clearly corresponded to the visible features of the diorama noticed by the students at the beginning of the activity. However, excluding the behaviour of the crows and the carrion, these "starter themes" were not revisited again. In contrast, when less conspicuous details of the diorama were noticed by the students, occasionally through teacher intervention, two themes, firstly, the yellow spots on the snow and secondly the footprints on the snow, which appeared more frequent in the students' reasoning. Besides these results, it was difficult to identify any further temporal patterns in these ten conversations. Temporal patterns in the conversation of families visiting an aquarium were investigated in a study of Ash et al. (2007). She found that patterns varied among families even though certain themes seem to direct the conversation (cf. starter themes). In a similar way to our study, certain elements of the exhibit in the aquarium were displayed prominently to encourage the initial conversation but the direction of the conversation was determined by the family (Ash et al. 2007).

When examining the relationship between the different themes in conversations, our results show that the *relations* established to (and within) the themes the yellow spots and the footprints, helped the students significantly to go further with the conversation and fill the gap of the identity of the carrion. Interestingly, the only group that recognized almost immediately the identity of the carrion had the longest and most diverse conversation. Other studies have also shown that recognizing an object as being familiar, was an important event in students' discourse in a museum visit and had a large impact on the outcome of the activity (Gilbert and Priest 1997). Importantly, the students' knowledge of anatomical details in relation to identifying the carrion, (i.e. the neck) showed a remarkable coincidence between groups. Moreover, as we showed in our previous study, the analysis of the students' relations made in the theme of the carrion, showed that students needed to make a variety of distinctions about the carrion as a particular object, before they could eventually decide what kind of animal it was (Piqueras et al. 2012). In a study of students' conversations in laboratory work Hamza and Wickman (2009) showed that the resolution of a specific task demanded linguistic investigations of labelling and distinguishing motivated by the particulars and contingencies of the problem. Regarding the theme of the carrion, where the main problem was identifying the animal, our results are in agreement with a variety of studies that have shown the important role of identification, naming, and labelling to cope with the different activities in the museum. In family conversations, for example, Tunnicliffe (2008) showed the importance of naming and recognizing parts of animals as very important components in their conversations when parent and children try to make sense at a dinosaur exhibit. Ash et al. (2007) found that naming (which she called characterizing talk), must occur before the formal science can begin (ecological talk). Similarly Allen (2002) found that, in families visiting an aquarium exhibition, the most frequent category of talk in families (perceptual talk) consisted of naming objects of the exhibit, identifying the features of the organisms, noticing parts of the exhibit and reading aloud exhibit texts. Overall, there is strong evidence that people need to learn how things and objects are named in particular practices and contexts (i.e. students in school activities or visitors in a free-choice learning environment) and, as stressed by Hamza and Wickman (2009), this *taxonomic interest* (Schwab 1978) has to be met in order to continue with more explanatory reasoning.

Teacher's intervention was important when noticing the footprints and the yellow spots in several groups when these details of the diorama were not commented spontaneously. This confirms our previous study (Piqueras et al. 2012), where we showed the important role of the teacher (or museum educator) when noticing gaps throughout short and concise interventions. For instance, by focusing their attention on specific details in the diorama, the students could engage in other themes and continue with the activity. The roll of the teacher and as mediator in the activity, scaffolding the students and providing opportunities of learning has been demonstrated in several studies (DeWitt 2012; Griffin 2012; Quistgaard 2010; Tal 2012). Similarly, debriefing activities of educational staff for visitors have been revealed as very effective providing further opportunities of exploration in exhibitions (Rowe and Kisiel 2012). In museum and other informal learning environments, several studies on family conversations have demonstrated the effect of parental interventions for scaffolding children's learning and inquiry skills in science (Ash 2003; Palmquist and Crowley 2007; Tare et al. 2011; Ash 2002; Siegel et al. 2007).

Our results also suggest that the possibility for the students to write down their own questions without restriction and discuss them afterwards in a small group is important for the development of a conversation that promotes learning. Moreover, facing their own questions resulted in the noticing of gaps in the encounter with the diorama, gaps which turned out to be crucial to determine the outcome of conversation. The generative role of questions in the students' conversations can be compared with a study about how families used questions at dioramas in Ash (2004). In her study, in which not only explicit questions are considered but also what is noticed (cf. gaps), she found that questioning invited rich family dialogue and generated the development of biological themes. In educational activities in museums, other studies have shown that intercalated open-ended questions in educational activities encourage students to reflect on the objects they see in the exhibition and hence become encouraged to discuss key concepts (Gilbert and Priest 1997; Quistgaard 2010).

References

Allen, S. (2002). Looking for learning in visitor talk: A methodological exploration. In G. Leinhardt, K. Crowley, & K. Knutson (Eds.), *Learning conversations in museums* (pp. 259–303). Mahwah: Lawrence Erlbaum Associates Publishers.

Ash, D. (2002). Negotiations of thematic conversations about biology. In G. Leinhardt, K. Crowley, & K. Knutson (Eds.), *Learning conversations in museums* (pp. 357–400). Mahwah: Lawrence Erlbaum Associates Publishers.

Ash, D. (2003). Dialogic inquiry in life science conversations of family groups in a museum. *Journal of Research in Science Teaching, 40*(2), 138–162.

Ash, D. (2004). How families use questions at dioramas: Ideas for exhibit design. *Curator: The Museum Journal, 47*(1), 84–100. https://doi.org/10.1111/j.2151-6952.2004.tb00367.x.

Ash, D., Crain, R., Brandt, C., et al. (2007). Talk, tools, and tensions: Observing biological talk over time. *International Journal of Science Education, 29*(12), 1581–1602.

DeWitt, J. (2012). Scaffolding students' post-visit learning from interactive exhibits. In E. Davidsson & A. Jakobsson (Eds.), *Understanding interactions at science centers and museums* (pp. 173–192). Rotterdam: SensePublishers.

Gilbert, J., & Priest, M. (1997). Models and discourse: A primary school science class visit to a museum. *Science Education, 81*(6), 749–762.

Griffin, J. (2012). In E. Davidsson & A. Jakobsson (Eds.)., *Understanding interactions at science centers and museums Exploring and scaffolding learning interactions between teachers, students and museum educators* (pp. 115–128). Dordrecht: SensePublishers.

Hamza, K. M., & Wickman, P. (2009). Beyond explanations: What else do students need to understand science? *Science Education, 93*(6), 1026–1049.

Kamcke, C., & Hutterer, R. (2015). History of dioramas. In S. D. Tunnicliffe & A. Scheersoi (Eds.), *Natural history dioramas: History, construction and educational role* (pp. 7–21). Dordrecht: Springer.

Kelly, G., McDonald, S., & Wickman, P. (2012). Science learning and epistemology. In B. J. Fraser, K. Tobin, & C. J. McRobbie (Eds.), *Second international handbook of science education* (Vol. 24, pp. 281–291). Dordrecht: Springer.

Palmquist, S., & Crowley, K. (2007). From teachers to testers: How parents talk to novice and expert children in a natural history museum. *Science Education, 91*(5), 783–804.

Piqueras, J., Hamza, K. M., & Edvall, S. (2008). The practical epistemologies in the museum: A study of students' learning in encounters with dioramas. *Journal of Museum Education, 33*(2), 153–164.

Piqueras, J., Wickman, P., & Hamza, K. (2012). Student Teachers' moment-to-moment reasoning and the development of discursive themes. An analysis of practical epistemologies in a natural history museum exhibit. In E. Davidsson & A. Jakobsson (Eds.), *Understanding interactions at science centers and museums* (pp. 79–96). Dordrecht: SensePublishers.

Quistgaard, N. (2010). Autentiske spørgsmål kan skabe ægte engageret dialog på naturhistoriske museer. *MONA, 3*, 49–76.

Rowe, S., & Kisiel, J. (2012). Family engagement at aquarium touch tanks. Exploring interactions and the potential for learning. In E. Davidsson & A. Jakobsson (Eds.), *Understanding interactions at science centers and museums* (pp. 63–77). Dordrecht: SensePublishers.

Scheersoi, A. (2015). Catching the Visitors' interest. In S. D. Tunnicliffe & A. Scheersoi (Eds.), *Natural history dioramas: History, construction and educational role* (pp. 145–159). Dordrecht: Springer.

Schwab, J. J. (1978). The nature of scientific knowledge as related to liberal education. In I. Westbury & N. J. Wilkof (Eds.), *Science, curriculum, and liberal education: Selected essays*. Chicago: University of Chicago Press.

Siegel, D. R., Esterly, J., Callanan, M. A., et al. (2007). Conversations about science across activities in Mexican-descent families. *International Journal of Science Education, 29*(12), 1447–1466.

Tal, T. (2012). Imitating the family visit. In E. Davidsson & A. Jakobsson (Eds.), *Understanding interactions at science centers and museums* (pp. 193–206). Dordrecht: SensePublishers.

Tare, M., French, J., Frazier, B. N., et al. (2011). Explanatory parent-child conversation predominates at an evolution exhibit. *Science Education, 95*(4), 720–744.

Tunnicliffe, D. S., & Scheersoi, A. (2015). Dioramas as important tools in biological education. In S. D. Tunnicliffe & A. Scheersoi (Eds.), *Natural history dioramas: History, construction and educational role* (pp. 133–143). Dordrecht: Springer.

Tunnicliffe, S. D. (2008). Conversations of family and primary school groups at robotic dinosaurs in a museum? What do they talk about? *Journal of Elementary Science Education, 20*(3), 17–33.

Tunnicliffe, S. D. (2015). Naming and narratives at natural history dioramas. In S. D. Tunnicliffe & A. Scheersoi (Eds.), *Natural history dioramas: History, construction and educational role* (pp. 161–186). Dordrecht: Springer.

Wickman, P. (2004). The practical epistemologies of the classroom: A study of laboratory work. *Science Education, 88*(3), 325–344.

Wickman, P., & Östman, L. (2002). Learning as discourse change: A sociocultural mechanism. *Science Education, 86*(5), 601–623. https://doi.org/10.1002/sce.10036.

Zimmerman, H. T., Reeve, S., & Bell, P. (2010). Family sense-making practices in science center conversations. *Science Education, 94*(3), 478–505.

Jesús Piqueras has a background in Ecology, but since 2004 he has been working in research and teaching in Science Education. His research focuses on Science teaching and learning in informal learning settings as museums, science center, botanical gardens and aquariums. Particularly, he has been interested in the study of learning process in socially shared practices in museum programs and activities designed for the school. Jesús is Senior Lecturer at the Department of Mathematics and Science Education, Stockholm University.

Karim Hamza is Senior Lecturer in Science Education at the Department of Mathematics and Science Education, Stockholm University. His research concerns the development of empirically grounded didactic tools or models that may be of use to science teachers in their day-to-day practice. Such "tools for teaching" may range from conceptual frameworks which help teachers plan, conduct, and evaluate their teaching, to more general hypotheses for how to reach certain purposes through different ways of teaching.

Susanna Edvall is a museum educator at the Swedish Museum of Natural History in Stockholm. During the last years, with her team in the museum, Susanna has developed a variety of school programs, activities and workshops that utilize the museums exhibitions as a tools and resource for inquiry based learning. During the last years, Susanna has participated in several museum research projects.

Chapter 14
The Use of Natural History Dioramas for Science Education

Michael J. Reiss

14.1 The Purposes of Science Education

There are a number of aims for science education (Reiss 2007) though these are often implicit. A frequent aim of many school science courses, and an aim supported by governments and industry, has been for them to provide a preparatory education for the small proportion of individuals who will become future scientists (in the commonly understood sense as employed professionals). This aim, though important economically, has been critiqued on democratic grounds (e.g. Millar and Osborne 1998). After all, what of the large majority of school students who will not become such scientists?

Another aim of science education, whether it takes place in schools or not, is to enable 'scientific literacy'. Although there has been a long-running debate as to the meaning of the term (e.g. Miller 1983), generally scientific literacy is seen as a vehicle to help citizens understand scientific issues. The basic notion is that science education should aim to enhance understanding of key ideas about the nature and practice of science as well as some of the central conclusions reached by science.

Many science courses in schools, colleges and universities hope that students will be able themselves to benefit materially from the science they have learnt. At its most straightforward this might be by entering or being in paid employment where the work that individuals undertake draws on what they have learnt of science. Although, as noted above, most school students do not enter such careers they too may still benefit individually from the science they have learnt whether inside or outside of school. For example, in most science courses, in countries around the world, it has long been accepted that one of the justifications for the inclusion of

M. J. Reiss (✉)
University College London, London, UK
e-mail: m.reiss@ucl.ac.uk

A. Scheersoi, S. D. Tunnicliffe (eds.), *Natural History Dioramas – Traditional Exhibits for Current Educational Themes*,
https://doi.org/10.1007/978-3-030-00175-9_14

certain topics is that knowledge and understanding of them can promote human health.

With John White, I have argued that that school education should equip every student to lead a life that is personally flourishing and to help others to do so, too (Reiss and White 2013). This aim applies to science education as much as to other subjects. Indeed, it is rather easier to see how school science might meet this aim than is the case for some other school subjects.

Whatever the agreed aims of science education, and whether we restrict ourselves to school-aged learners or to others, my contention is that natural history dioramas have a small but significant role to play in meeting the aims of science education. In this, the concluding chapter to this book, I examine the key arguments and conclusions of each of the other chapters, concentrating on the extent to which natural history dioramas are indeed known to help meet the aims of science education.

14.2 Fashion and Dioramas

Natural history dioramas have a long history in many countries; indeed, as discussed by Rogers, Shreckengast and Dorfman in Chap. 2, their origins can be traced to the late eighteenth century presentation of taxidermy specimens by Charles Willson Peale within a recreated, natural environment. Peale soon abandoned this new approach, probably because the uncased animals were subject to damage by insects, patrons and sunlight (problems that remain to this day), but others took up the reins. Important among these, though overlooked till now in major publications detailing the early history of dioramas and habitat groups in the United States, was another taxidermist – James A. Hurst. From the 1850s, Hurst used his own extensive collection of birds and beasts of North America to produce stereoscopic views. These offered the public an innovative way to view a representation of the natural world from the comfort of their homes. Whether these paintings, combined with a taxidermy mount and recreated foreground, were permanent exhibits is unknown.

Despite this long history, dioramas went out of fashion towards the end of the twentieth century in the USA and in many countries. Fashion is a funny thing. One is reminded of how Andrew Lloyd Webber's grandmother responded "I will not have Victorian junk in my flat" when her grandson begged her in the early 1960s to lend him £50 so that he could buy Lord Leighton's *Flaming June* – now known as the Mona Lisa of the southern hemisphere. (Years later Webber offered £6m for it but was turned down.) In the case of dioramas, they seem to have been become seen as old-fashioned compared with the glitzy allure of interactive exhibits and other 'modern' methods of display. Their problem was precisely that they were objects for education. A museum isn't likely to destroy what it considers to be historical objects of cultural importance however fuddy-duddy they look but will happily consign to skip the work of craftsmen, who labored for hundreds of hours to produce a single diorama for educational purposes.

It's too early to say that the tide has turned but dioramas are making something of a comeback, despite their cost. They are often much loved by visitors of all ages and, as the other chapters in this book discuss, have considerable educational potential.

14.3 Developing and Maintaining Dioramas

Our relationship with wolves is a complicated one. While some see wolves as a persecuted species, now rare or extinct in many countries, so that we have conservation responsibilities to them, others see them as pests, others as dangerous, others as the rather bored inhabitants found in wildlife sanctuaries, others as the ancestor of 'Man's best friend' and others as more general reservoirs of cultural resonance – e.g. the story of Little Red Riding Hood (Reiss and Tunnicliffe 2011). All of these possible readings are open to a viewer seeing an isolated wolf but, as Karen Wonders (1993) points out, a natural history diorama encourages some readings at the expense of others.

In Chap. 3, Andrew Kitchener tells the story of his role in the development of a new diorama, wolves and all, that deals with the biological history of Scotland since the end of the last Ice Age. He points out that many dioramas show animals just standing or looking back at the observer. He wanted to show animals displaying dramatic natural behaviours and interacting with each other, in ways that highlighted key adaptations, behaviours or interactions between species. He also wanted to show not just the big changes from tundra at the end of the last Ice Age to the peak climax forests of up until about 5000 years ago (when Neolithic farmers began to change the natural landscape dramatically and permanently), but also changes on a seasonal and daily timescale.

From an educational perspective, one of the difficult decisions was whether to show people or not. In the end, people were excluded partly because the Mesolithic human story is told in an adjacent gallery and partly because it was felt that there was no way of knowing what clothes people would have worn at the time. However, there is a very small snowman – though the visitor has to look from the right vantage point to see him. Evaluation to date has been somewhat limited (Tunnicliffe 2005), though from the two occasions I have visited the diorama I can vouch for the affection in which it is evidently held, especially among younger visitors.

In Chap. 4, Eirik Granqvist reflects on his experiences of the construction of dioramas in the Zoological Museum of the University of Helsinki, beginning with one that shows a fight between elk (*Alces alces*) in the autumn rut. He researched their behaviour and spent 3 years before he found the specimens that he wanted, which he shot himself. Comparable care was taken over the other organisms, the painting of the background, the lighting, the viewing angle and the size of the window.

Granqvist states bluntly that the greatest danger to dioramas comes from museum staff who are not competent or knowledgeable in diorama construction and maintenance. After that, the chief problem is insects – as also noted by Kitchener in Chap. 3 – since clothes moths came close to running the diorama only a few years after it had opened. There is a certain poignancy to Granqvist's comments about museum staff as his own dioramas in the Zoological Museum of the University of Helsinki no longer exist, having been dismantled.

14.4 The Educational Significance of Natural History Dioramas

As becomes clear on reading the chapters in this book, despite some impressive exceptions, many of them undertaken either by Sue Dale Tunnicliffe herself or by colleagues or doctoral students of hers, the educational significance of natural history dioramas remains considerably underexplored.

An example of this lack of exploration is provided in Chap. 5 by Rainer Hutterer and Till Töpfer. Despite titling their chapter 'Dioramas of Marine Bird Colonies: History, Design, and Educational Importance', the material on educational importance is really about educational potential. For a case study, they focus on the Bird Rock on Bear Island diorama, found in the Museum Koenig in Bonn and constructed from 1912 to 1933. Bear Island lies between Spitsbergen and the Norwegian North Cape. Koenig made a number of visits to the Spitsbergen archipelago and was fascinated that its bird fauna was still relatively unexplored. Scientific publications, authored by Koenig and others who accompanied him, resulted for Bear Island and Spitsbergen.

The bird rock diorama is impressive in its scale: it stands some 5 m and, despite a number of changes to its lighting over the years, it is still very much in its original form and benefited from a thorough renovation in 2006. In all, there are 98 birds belonging to a total of nine species.

Hutterer and Töpfer discuss the educational potential of the diorama. Clearly the visitor can learn much about bird morphology, including adaptations for flight and the existence of two plumage morphs in the common guillemot (*Uria aalge*), as well as about seabird diversity, breeding ecology, egg characteristics, feeding ecology and the importance of rock islands for bird conservation, not to mention expedition history. Hutterer and Töpfer suggest that additions could be made to enhance the educational value of the diorama; for example, there could be an acoustic background with sound recordings from a bird rock colony on Bear Island, or live pictures from a remote webcam from a seabird colony somewhere in Europe.

A different approach is adopted in 'miniature dioramas', of which a number of botanical examples exist. In Chap. 6, Kathrin Grotz discusses the sixteen small vegetation dioramas (with a scale of 1:10 or 1:20) in the Botanical Museum in Berlin. In passing we can note the extraordinary amount of time and effort that was spent on constructing these dioramas: for example, the three alder trees standing at the 'Lake of the North German lowland' required 12,000 custom-made miniature sheets of painted copper sheet. However, while the aesthetic and emotional value of the dioramas is undisputed, most visitors to the Botanical Museum find it difficult to understand what they convey. Grotz discusses how these dioramas can be updated to make them more intelligible to visitors.

In Chap. 7, Alexandra Moormann and Charlène Bélanger point out that since they were first produced, natural history dioramas have always been intended to trigger learning among visitors about biological topics such as animal behavior and ecosystems. In their work, they look at dioramas as simplified models of natural ecosystems and seek to answer two questions: To what extent can dioramas be considered scientific models? How can dioramas promote model-based learning?

Moormann and Bélanger point out that before the appearance of dioramas, natural science museums were mainly concerned with the exhibition of objects from their collections, organised according to the structure dictated by the understanding of the natural order that was dominant at that time in the scientific community. By seeing dioramas as models, we can appreciate that dioramas are not meant to show reality 'as it is' since this is not the function of a model. Rather, a model inevitably embodies a reduction and particular refraction of reality. Accordingly, a diorama, like a model, will present different features depending on intentions of their creator.

Furthermore, models, and hence dioramas, can be seen as existing not only in the form of tangible three-dimensional objects but also as the mental representations of complex ideas or systems (Gilbert and Priest 1997; Clement 2000). Despite this, as Moormann and Bélanger note, research in science education about models has shown that students often understand models to be replicas of reality rather than as instruments within an epistemological process. This leads to Moormann and Bélanger's second question: How can dioramas promote model-based learning?

The process starts from the act of museum designers and educators who transform scientists' consensus models into museographic target models. As visitors interact with dioramas taken as museographic models, mental models can gradually develop. Moormann and Bélanger have started exploring the learning about models that occurs during a school visit at the Museum für Naturkunde Berlin. Using concept maps and group discussions, they have been able to describe different forms of conceptual change with regard to students' understandings. Using the museum-derived framework, they plan to look at the gradual transformation of the students' mental models about natural ecosystems through repeated and varied encounters with dioramas.

Michael May's and Marianne Achiam's interest in Chap. 8 is on the specific mechanisms that enable the educational potential of dioramas to be realised. As they put it: "museum practitioners and museum researchers know that dioramas work, we just don't know how they work". They begin by noting that designers and curators tend to conceptualise museum visits as the *communication* of design intentions to visitors, mediated by exhibits, labels, explanatory signs and narratives. However, from the museum visitor's point of view, the experience can best be described as a *meeting* with artefacts or natural objects on display. As we all know, meetings can take place with little worthwhile communication occurring.

The inquiry aspect of a museum visit arises when visitors are confronted with exhibited objects they cannot immediately identify or with situations they do not understand. This can set in motion a sequence of actions such as looking for information on labels or explanatory signs next to a diorama. As a naturalistic or pseudo-naturalistic display of objects (e.g. humans, animals, plants, machines) in their natural or cultural environment, dioramas and their interpretation can appear straightforward, but, argue May and Marianne, complexity hides behind the naturalistic surface.

May and Achiam draw on the basic Gestalt principles formulated by Max Wertheimer almost 100 years ago. According to the *proximity* principle we experience objects that are spatially close as meaningfully grouped together, and according to the *similarity* principle we experience objects that are visually similar in shape, size or colour as meaningfully grouped together. Equally, we divide views into *foregrounded figures* and the *background*. Furthermore, we extrapolate from what we see, attempting to give meaning to what is before us, as indicated by the principle of *continuation*, whether spatially or temporally.

These principles are important for how we understand dioramas. A moment in time showing, for example, the tiger diorama from the Natural History Museum in Helsinki (Chapter 8, Figure 8.2) is read by us as a tiger attacking a deer (or a deer being attacked by a tiger, depending on our literal and metaphorical point of view) even though such a phrase indicates a considerable passage of time and a degree of intentionality. In other words, we extrapolate from what is in front of us; we generate a narrative to make sense of what we see, imagining both a past and a future rather than restricting ourselves to the present before us. Of course, such behaviour is honed by our evolutionary past (Abrahams and Reiss 2012). Early humans who failed to presume that a large predator in mid-air close by them might best be avoided did less well than those who did so presume.

There are educational risks in all this. As May and Achiam point out, given that the learning potential of the diorama arises from the imaginative richness of the content it creates by placing artefacts and natural objects within a naturalistic scene, one potential problem is that the naturalistic articulation of the diorama and its power to generate stories creates an apparent realism that could entail false inferences. This suggests an important role for a teacher in helping visitors understand the relationships between the diorama in front of them, their readings of it and, returning to Moormann's and Bélanger's point in Chap. 7, the reality that the diorama is attempting to model.

In Chap. 9, Edward Mifsud proposes that Engeström's (1999) Activity Theory can profitably be used to illuminate what is going on when a visitor attempts to interpret a museum diorama. Engeström's Activity Theory itself derives from the classic early work of the Russian psychologists Vygotsky, Leont'ev and Luria. One of the key points of Activity Theory is that it takes seriously the fact that the learning of individuals does not take place in isolation: learning is embedded in social relationships, so that we need to consider each individual, the objects of their learning and the social context in which learning takes place.

Mifsud applied his model to to data obtained from a *subject*, a 9-year-old boy, Jeremy, in the fifth year of primary school, who was observing a sand dune diorama in Malta (Chap. 9, Fig. 9.4), the *artefact*. The *group* is his class, particularly the pupils with whom he was experiencing the diorama. The diorama shows various birds, some reeds and a prominent Maltese boat resting on a bed of sand. Jeremy observes the diorama, which acts as a mediating tool to aid in his interpretation of the sand dune habitat it represents. He does this in the company of his peers, which may influence the way he sees the diorama and what he notices.

Mifsud uses drawings to help reveal the mental models of those at the diorama. Comparing the expressed models with the original model helps to reveal the importance of *previous knowledge*. Malta is, of course, a relatively small island and so Jeremy was previously familiar with seaside habitats.

While the strength of the model lies in how it links together the elements involved in the interpretation of an artefact (in this case a diorama), Mifsud lists a number of limitations, though some of these seem to have more to do with his model than with the use of Activity Theory.

In Chap. 10, Debra McGregor and Jennifer Gadd move the focus to how beginning teachers can be supported to help children understand the work of natural history scientists. Instead of undertaking work at existing professionally made dioramas – the standard approach in educational research on dioramas – McGregor and Gadd got beginning teachers to make their own small dioramas.

The beginning teachers were in the second year of a Bachelor of Arts (Educational Studies) undergraduate degree course. A total of 71 such students were split into three (successive) teaching groups and took part in a specially designed workshop, in which they mostly worked in triads. Each group of three students was given a card with the name of one of the natural history scientists named in the 'notes' section of the latest science National Curriculum in England: Mary Anning, David Attenborough, Rachel Carson, Charles Darwin and Jane Goodall. They were also given a shoe box within which to create a diorama, and a range of everyday modelling materials such as paper, card, tissue, coloured cellophane, paints, hot glue guns and cutting knives. The students were asked not to divulge which scientist they had been allocated; their task was to 'reveal' their scientist through their diorama. Finally, they were asked to create their own success criteria for their displays, taking into account what they might expect a primary school child to achieve during the activity.

As they constructed their dioramas, the students were provided with a tablet computer connected to the internet to help them research their scientist. The tutors (Debra McGregor and Jennifer Gadd) moved from group to group collecting field notes and taking photographs to document the artefact construction process. Once the building-dioramas phase was complete the beginning teachers took the time to visit and review the other dioramas, as if at a local museum reviewing the display boxes.

Unsurprisingly, the evaluation revealed that the students enjoyed the activity and most of them indicated that they learned much more about the lives and work of natural history scientists. The dioramas illustrated how the groups focused on differing aspects of the scientists' work. For example, and just with reference to Charles Darwin, one diorama highlighted his conclusions about the evolution of humans from (other) primates; another focused on his work, symbolised by an enlarged magnifying lens, on the beaks of finches on the Galapagos Islands; a third looked at variation in the general morphology of the finches. Darwin was well known to the students; at the other extreme, not a single one of the 71 had heard of Rachel Carson. From knowing nothing about her, the students came to realise that she possessed keen observational skills that informed her work on the causal relationship between pollution and damage to the reproductive capacity of organisms.

Interesting too were the students' views as to why dioramas were appropriate to use for learning. About half the students recognised a wide variety of learning processes involved in the construction and review processes of making the dioramas. They commented on the way that the collaborative learning processes were fun and involved discussions, exchanges of ideas and negotiations about what to include and how to make the displays. They realised that researching for relevant information and considering ways to use resource materials creatively to represent relevant ideas that illustrated and summarised understandings all involved critical and in-depth thinking. Some thought that using dioramas for learning could promote motor and design skills.

Cristina Trowbridge is a senior manager of professional development at the American Museum of Natural History. Initially she did not use dioramas much in her work but she wanted to help teachers harness the potential for interactive learning inherent in the museum setting. In 2009, she met Amy Chase-Gulden who taught her Visual Thinking Strategy (VTS); this changed her way of interacting with the dioramas and in turn how she worked with teachers and students.

Visual Thinking Strategy (VTS) is an interpretive method that is utilised more in art museums than in natural history museums. Traditionally, VTS is used in a group setting while looking at art and is facilitated by three guiding questions: *What is going on in this picture?*, *What do you see that makes you say that?* and *What more can we find?* VTS is intended to create a non-judgemental, facilitating environment that removes the risk of failure, welcomes plurality of perception, evokes curiosity and invites engagement.

Trowbridge added a sketching component to VTS, which proved to be a successful modification, and in Chap. 11 relates some of her experiences. She found that VTS with sketching helped teachers and students to engage in careful observation, an important component of science. It had other benefits too. As one first year teacher working with predominately second language learners in an Earth Science high school class put it:

> Like meditation, it required focus, but was enjoyable. It was meaningful to pay attention to small details that would normally go overlooked. It created an allotment of time that was reserved for silence and peace. It forced us to have a break from the busy schedule of teaching to relax and focus on detail. Specifically since induction had a lot of discussion on the challenges of teaching, and was the ending of a stressful workweek. It was a great stress releaser that kept my mind active. I've thought about going to the museum on the weekend to draw some of the dioramas.

Additionally, the approach modelled other key aspects of the nature of science, namely inference and the formulation of hypotheses. Trowbridge concluded that the use of VTS and sketching with natural history dioramas offers new science teachers a readily replicable process for engaging students in direct observation of and inquiry about representations of natural phenomena.

In Chap. 12, Sue Dale Tunnicliffe, Rebecca Gazey and Eirini Gkouskou point out that in many early years settings, the educators, be they carers, primary/elementary teachers or chaperones on school field trips, are not confident in their own understanding when teaching physical science. Natural history dioramas therefore offer the promise of providing a more accessible route into physical science concepts, e.g. forces, than if such concepts are approached as they usually are, without the mediation of a biological context. Indeed, visitors at natural history dioramas often comment about such elements of physical science as the movement of animals. Other physical science concepts that may be facilitated by natural history dioramas include those of balance, the reflection of light, centre of gravity and heat loss.

In a workshop undertaken with two 11-year-old boys at the Powell-Cotton Museum, the boys were asked, after viewing the Watering Hole diorama (Chap. 12, Fig. 12.3), to make a four-legged animal that could stand upright, using modeling clay (for the body) and cocktail sticks (for the legs). One boy immediately made a flat (horizontal) cuboid and fixed four legs, one at each corner of the body. The other boy decided to make an animal with a vertical cylindrical body. It was not easy to get this animal to stand up. He decided to reorient the 'body' so that he had a rectangular one that lay horizontally. Then he fixed the kegs together in the middle of the underside of the 'body'. Eventually he decided to try positioning the legs at the four corners and was pleased to find that this produced a stable model. The boys were invited to stand their 'animals' on a piece of cards which acted as a 'wobble' board and to investigate for how long their animal stood as more and more backward and forward movements of the board were made. They found that, by having the legs not coming down vertically from the body but splayed out, the animals were more stable. The boys were asked to add a neck and a head on their body and then show how the giraffe-like model animal could drink. By simulating it visiting The Watering Hole, they found that the animal toppled over until they had increased the distance between the bottom of the legs. They remembered they had learnt about forces in school science but said that that had not related to anything in their everyday world.

On returning to the diorama, the boys added these science ideas of balance, stability and centres of gravity to their interpretation of the diorama. One of the boys reported that "the giraffe starts bending her legs to get her head closer to the ground. I can see stability in the animals. The legs support the position of the head. Each part of the body supports because, for example, one leg of the giraffe cannot work without the other legs".

Finally, in Chap. 13, Jesús Piqueras, Karim Hamza and Susanna Edvall present an analysis of student teachers' moment-by-moment learning during a teaching activity at a diorama at the Swedish Museum of Natural History in Stockholm. The diorama showed a female white-tailed eagle eating the remains of a roe deer in a snowy landscape in the presence of crows (Chap. 13, Fig. 13.1). An interview with the curator showed that the main intention of the diorama was to show the co-operative behavior of the crows as they attempt to steal food from the eagle. Another curatorial aim was to challenge the curiosity and imagination of the visitors by placing additional details and clues in the main scene. Thus, there are yellow spots and footprints interspaced on the snow, resulting from a presumptive fox that has visited the carrion before the eagle and the crows; indeed, the head of the roe deer is missing (a common behavior of foxes is to take away the head of the animals they predate). The diorama was inspired by a real event shown in a short video sequence on a monitor screen close to the diorama. However, this resource was kept hidden from the student teachers during the activity.

The ten student conversations at the diorama lasted an average of 12 min. Piqueras, Hamza and Edvall state that, to their surprise, the conversations were rarely related to the curatorial intention of the diorama but this is, I think, not unexpected for two reasons. First, few visitors, even student teachers, are likely to know anything about the way in which crows attempt to steal food from eagles whereas they are likely to know at least something about the way in which eagles eat carrion. Secondly, the decision was made not to allow the student teachers to view the video sequence on the nearby monitor screen, thus requiring the students to construct a narrative entirely from their own knowledge and presumptions.

14.5 Conclusions

Having read the chapters in this book, I have come to three main conclusions.

The first, one that I share with all the authors, is that natural history dioramas have tremendous educational potential. As has often been noted, such dioramas are engaging to many visitors, whatever their age and prior knowledge and experience. Natural history dioramas enable visitors to construct narratives about what is happening.

The second, somewhat less positive, conclusion is that there remains a paucity of evidence as to the extent and types of learning that natural history dioramas afford. This, of course, is a spur to further research.

The third conclusion is that there exists quite a wide range of research tools for gathering and analysing data about the learning that takes place at dioramas. Some will welcome this pluralism; others will wonder whether it tells us more about those undertaking such research than anything else.

References

Abrahams, I., & Reiss, M. (2012). Evolution. In P. Jarvis & M. Watts (Eds.), *The Routledge international handbook of learning* (pp. 411–418). Abingdon: Routledge.

Clement, J. (2000). Model based learning as a key research area for science education. *International Journal of Science Education, 22*, 1041–1053.

Gilbert, J. K., & Priest, M. (1997). Models and discourse: A primary school science class visit to a museum. *Science Education, 81*, 749–762.

Millar, R., & Osborne, J. (Eds.). (1998). *Beyond 2000: Science education for the future*. London: Nuffield Foundation.

Miller, J. (1983). Scientific literacy: A conceptual and empirical review. *Daedalus, 112*(2), 29–48.

Reiss, M. J. (2007). What should be the aim(s) of school science education? In D. Corrigan, J. Dillon, & R. Gunstone (Eds.), *The re-emergence of values in science education* (pp. 13–28). Rotterdam: Sense.

Reiss, M. J., & Tunnicliffe, S. D. (2011). Dioramas as depictions of reality and opportunities for learning in biology. *Curator, 54*, 447–459.

Reiss, M. J., & White, J. (2013). *An aims-based curriculum: The significance of human flourishing for schools*. London: IOE Press.

Tunnicliffe, S. D. (2005). What do dioramas tell visitors? A study of the history of the forests diorama at the Museum of Scotland. *Current Trends in Audience Research and Evaluation, 18*, 23–31.

Wonders, K. (1993). *Habitat dioramas: Illusions of wilderness in Museums of Natural History*. Uppsala: Uppsala University Press.

Michael J. Reiss is Professor of Science Education at the UCL Institute of Education, University College London, Honorary Fellow of the British Science Association, Visiting Professor at the Universities of Kiel, York and the Royal Veterinary College, Honorary Fellow of the College of Teachers, Docent at the University of Helsinki and a Fellow of the Academy of Social Sciences. Books of his include: Barmania, S. & Reiss, M. J. (2018) Islam and Health Policies Related to HIV Prevention in Malaysia, Springer; Reiss, M. J. & White, J. (2013) *An Aims-based Curriculum*, IOE Press; Jones, A., McKim, A. & Reiss, M. (Eds) (2010) *Ethics in the Science and Technology Classroom: A New Approach to Teaching and Learning*, Sense; Jones, L. & Reiss, M. J. (Eds) (2007). *Teaching about Scientific Origins: Taking Account of Creationism*, Peter Lang; Braund, M. & Reiss, M. J. (Eds) (2004) *Learning Science Outside the Classroom*, RoutledgeFalmer; Levinson, R. & Reiss, M. J. (Eds) (2003) *Key Issues in Bioethics: A Guide for Teachers*, RoutledgeFalmer; and Reiss, M. J. (2000) *Understanding Science Lessons: Five Years of Science Teaching*, Open University Press.

The manufacturer's authorised representative in the EU is Springer Nature Customer Service Centre GmbH, Europaplatz 3, 69115 Heidelberg, Germany. If you have any concerns regarding our products, please contact ProductSafety@springernature.com

Printed and bound by CPI Group (UK) Ltd, Croydon, CR0 4YY
30/09/2026
02225505-0006